The Texas Hill Country

THE TEXAS HILL COUNTRY

A Food and Wine Lover's Paradise

Second Edition

by Terry Thompson-Anderson

Photography by Sandy Wilson

With a Foreword by Todd Staples
Commissioner, Texas Department of Agriculture

 Shearer Publishing
Fredericksburg, Texas

Text copyright © 2008 and 2010
Terry Thompson-Anderson

Photographs copyright © 2008 and 2010
Sandy Wilson

Shearer Publishing
406 Post Oak Road
Fredericksburg, Texas 78624
Toll-free: 800-458-3808
Fax: 830-997-9752
www.shearerpub.com

Library of Congress Cataloging-in-Publication Data

Thompson-Anderson, Terry, 1946–
   The Texas hill country : a food and wine lover's paradise / by Terry Thompson-Anderson ; photography by Sandy Wilson ; with a foreword by Todd Staples.
      p. cm.
   Includes index.
   ISBN 978-0-940672-80-2
   1. Restaurants—Texas—Directories.
   2. Farms—Texas—Directories. 3. Cookery, Texas.
   4. Cookery, American—Southwestern style. I. Title.
   TX907.3.T4.T53 2008
   647.95764—dc22

                                    2008019891

ISBN 978-0940672-80-2

The recipe on page 74 was adapted with permission from The Pastry Queen: Royally Good Recipes from the Texas Hill Country's Rather Sweet Bakery & Cafe, Ten Speed Press, 2004.

Information from Web sites of the businesses listed in this book has been used with the permission of their respective owners.

First Edition, 2008
Second Edition, 2010

                              10 9 8 7 6 5 4 3 2

Production by Phoenix Offset

Printed in China

A Note to the Reader

This book is not meant to be comprehensive, but rather to provide a cross section of the food- and beverage-related businesses in the Texas Hill Country. We encourage you to visit and enjoy the many other venues the region has to offer!

The venues featured here are arranged by town, with the towns listed alphabetically. You can plan a trip of several days, visiting the many attractions found in several towns, or just single out a particular town to visit.

Although the entries are based on the most up-to-date information available and were verified prior to publication, Shearer Publishing does not guarantee or warrant the accuracy of any information in this product and will not be liable in any respect for any errors or omissions. We suggest that you contact the venue directly by phone or check its Web site before planning your trip.

Contents

Foreword

Because it serves as such a wonderful reminder of the value of sustainable agriculture practices, I invite you to enjoy Terry Thompson-Anderson's book, *The Texas Hill Country: A Food and Wine Lover's Paradise.*

At the Texas Department of Agriculture, we work hard to help those who rely on the fruits of the land: the peach, apple, olive, and pecan growers, the cheese and sausage producers, and the grape growers and winemakers. Texas offers an abundance of natural resources, and the Texas Hill Country in particular is a region rich with the treasures and rewards of plentiful agriculture. Top-notch products resulting from the tireless efforts of Texas producers are celebrated within these beautiful pages.

The food on our tables is so closely connected to the people and communities that surround us. This book introduces us to Hill Country farmers and growers who remain steadfast in their pioneering spirit and work ethic.

The landscape of the Texas Hill Country is an inspiring and breathtaking sight, from its sparkling, spring-fed, meandering creeks to its rugged terrain and rolling hills. Within the pages of this book you will see firsthand the true heart and soul of the region and all the resources it has to offer. You'll come face to face with the proud pioneering spirit for which hardworking Texans are known. Witness the contributions these Texans make on a daily basis, not only through their continued efforts to nourish our families with quality products, but also with their appreciation, dedication, and conservation of our beloved land.

Agriculture touches each of our lives and is a dynamic economic engine that powers our state's economy. Texas agriculture generates more than $85 billion throughout the Lone Star State, about 10 percent of the gross state product. In addition, agriculture provides a job for one out of every seven working Texans.

With the help of this book, discover the highlights that make the Texas Hill Country a national treasure. Savor the moments at appetizing area restaurants, bakeries, and herb farms. Visit award-winning wineries, tearooms, and apple orchards. Taste the rich delicacies of organic produce and perfectly aged cheese, and don't forget to take in the aroma of fresh Texas Hill Country flowers.

GO TEXAN and delight in the bounty of Texas agriculture. In Texas it's part of all of us.

Todd Staples
Commissioner, Texas Department of Agriculture

Introduction

The Texas Hill Country, a region encompassing all or part of eighteen counties in the geographical center of Texas, has been described as being somewhat like Jackson Pollock on a rampage. Fields of riotous color cover the countryside each spring when bluebonnets, Indian paintbrushes, and dozens of other wildflowers burst into bloom, announcing the end of winter in waving fields of many hues. Technicolor sunsets, glowing against brilliant cerulean skies, slip over purple-hued mountain ranges on the distant horizon. And speaking of horizons, we really have horizons—horizons that you can actually see in any direction.

The topography and venues of the Hill Country vary somewhat dramatically from region to region. The northwestern section has topography that ranges from the granite-strewn hills and spring-fed rivers around Llano to the fertile river-bottom pecan country of San Saba and Brady and acres of lush vineyards in the newly established grape-growing region around Mason. In the far northwestern reaches of the Hill Country around Menard the region is sparsely populated with a more pronounced look of only scattered vegetation and a rockier terrain. The area, known as the jumping-off point to the craggy Texas High Plains, contributed vastly to the westward expansion of the United States.

The northeastern region of the Hill Country includes some of the most popular recreational areas in the state. This is the part of the Hill Country where water sports abound on the chain of Highland Lakes, formed from the Colorado River, beginning with Lake Buchanan's headwaters at Tow (rhymes with "wow") and continuing through Inks Lake, Lake LBJ, Lake Marble Falls, and Lake Travis. The topography of the region is a mix of rolling hills in the eastern portion to craggy ranchland with huge granite outcroppings and limestone cliffs in the central and western portions. The land, however, is fertile, supporting an abundance of agricultural endeavors. Many scenic vistas of the limestone and granite cliffs can be seen best from vantage points on the water in the Highland Lakes.

Traveling through the fertile river-valley region of the southeastern Hill Country, the visitor finds a greater population density around New Braunfels and San Marcos, easing into smaller towns with a more laid-back ambiance like Wimberley and Driftwood. Heading west, you'll find Dripping Springs, often referred to as a "bedroom community of Austin," then on to Johnson City, Blanco, and Stonewall, towns with a definite feel of the old Texas cowboy life. The land is composed of rolling hills and vast stretches of open ranch and farm land. The area from Stonewall to Fredericksburg is home to one of the state's biggest agricultural endeavors, peach and blackberry farming.

The southwestern portion of the Hill Country, known as the Edwards Plateau region, was formed millions of years ago by the uplifting of a onetime seabed, creating a unique geology. The region is home to some of the oldest rocks in Texas. Part of the region, around Fredericksburg, is characterized by large granite outcroppings and domes, such as the popular tourist destination Enchanted Rock. It is a region of many springs, stony hills, and steep canyons. Oak and cedar trees are the predominant vegetation. This was a region of great challenge for the early settlers with its coarse-textured sandy soil, the result of granite weathering over thousands of years. In this section of the Hill Country the visitor finds the German-settled cities of Fredericksburg, Comfort, and Kerrville, which have retained a pronounced German flavor. Fredericksburg has several annual events celebrating its German heritage. Shopping and good restaurants are hallmarks of each of the cities. Between Kerrville and Bandera you'll find the sleepy little town of Medina tucked among gigan-

tic oaks in close proximity to the meandering Medina River. In Medina you'll find an oasis of apple orchards. Bandera is cowboy country, home to dozens of working cattle, sheep, and goat ranches. Shopping in Bandera is an adventure in cowboy culture, yet the city boasts some thoroughly modern, upscale eateries. In the far western reaches of the southwestern Hill Country you'll find the quaint town of Leakey (pronounced LAY-key), located on the spring-fed Frio River and surrounded by deep canyons and hills with elevations of up 2,400 feet. Junction, originally founded in 1876, is known as "The Land of Living Waters" because there is more flowing water in the city and surrounding Kimble County than any other county in the state! The city today thrives as a shipping and marketing center for the county's wool, mohair, livestock, pecan, and grain production. The area around Junction is also known as one of the state's major deer-hunting areas, providing a winter economic windfall to many area businesses.

Those of us who are privileged to have stewardship of our little corners of this awe-inspiring region are fierce in our dedication to preserve both the unique culture and the land. The original pioneer settlers of our region and those who followed closely on their heels braved great odds just to get here, then they had to fight for the right to stay here, often suffering grave losses in the building of the Hill Country legacy. They were farmers and ranchers and sausage makers, bakers, brewers, and vintners. It's a place where the word "roots" means something, where finding fourth and fifth generations in our small towns is the rule rather than the exception.

We see our challenge today as having the wisdom to guide our small towns, those hidden treasures tucked among the hills, rivers, and lakes, in the inevitable switch from agricultural-based to tourist-based economies, balancing the tide of progress to allow room for both. To preserve and guide the entrepreneurial spirit established here by the pioneers. To welcome newcomers who have the potential to indelibly change our way of life, hoping that they will instead remain mindful of the reasons why they came here and help us to sustain that way of life.

The pioneer spirit is still very much alive in the Texas Hill Country. The newest group of pioneers to set down roots are the lavender growers. The willingness of the growers to face the possible hardships and the challenges of growing lavender in this region has changed the face of the community—both physically and economically through the landscape's aromatic and beautiful blossoms and the great increase in tourism.

Today there are more than fifteen lavender farms stretching from Fredericksburg to Johnson City, Blanco, and Wimberley. The city of Blanco, which calls itself the Lavender Capital of Texas, hosts an annual Lavender Festival the second weekend in June. The festival features tours of area lavender farms, each with special treats awaiting guests and cut-your-own lavender, speakers on a variety of lavender-related topics, and a Lavender Market on the grounds of Blanco's Courthouse Square. Lavender is becoming a big business.

The entrepreneurial spirit has also turned Mason County, with its slightly acidic Hickory Sands, into a prime region for growing wine grapes. The area has a unique microclimate with ideal daytime temperatures and strong breezes that help to prevent problems with fungus and mold on grapevines. The Hickory Sands Aquifer also provides an ample supply of good, clean water with a nearly neutral pH for irrigation. To protect the vines from damage by the hard spring freezes that plague the area, the ingenious growers of Mason County have devised a unique way to use a drip irrigation system to cover the vines with ice during a freeze. When the freeze is over and the ice thaws, the little buds grow on their merry way to becoming grapes.

So how are we doing in the Texas Hill Country? Well, pretty spectacular, we think! Hill Country folks are creating ways to turn that tourist-based economy into what is often referred to as "agricultural tourism." Visitors are coming to the Hill Country to follow the lavender trails, pick peaches and apples, buy fresh-pressed olive oil, and visit the wineries. In fact, the Hill Country was listed in a recent survey by Internet travel giant Orbitz as second only to the Napa Valley as an agricultural tourism destination!

Pick a road—any road, a county road, a ranch road. Some are paved, some aren't. Some are sectioned off by cattle guards, with vast expanses of unfenced land in between where livestock roams free. Some are bordered by ancient barbed-wire fences constructed with hand-cut cedar posts. Some meander past sprawling ranches with tall, deer-proof fencing and grand entryways fashioned from native stone and hand-wrought iron artwork. Chances are good that somewhere down that undulating ribbon of road, you'll come upon some remarkable Texas Hill Country food finds: a farm stand selling fresh, homegrown produce; a goat farmer selling hand-crafted goat cheeses; ranchers selling sausages made from game meats; a brewery producing fine beer in the European style of the region's founders; a winery creating world-class wines from grapes grown in Texas; honey processors selling honey from bees who feast on the region's largesse of richly hued wildflowers; peach orchards with roadside stands selling fresh peaches, peach ice cream, peach cobbler, and anything else that can be made from a peach; apple orchards where you can purchase many varieties of fresh apples, apple cider, and apple pies as big as washtubs.

Perhaps you'll find a berry farm where you can pick your own berries; pecan farms where thousands of towering pecan trees produce huge crops of delicious pecans each fall for sale in farm stores, where you might also find sinfully rich homemade pecan pies; fields of lavender in riotous purple bloom; the best barbecue anywhere, often served from roadside stands; free-roaming herds of Texas Longhorn cattle grazing on pesticide-free natural grasses; unique and often quirky cafés serving the kinds of foods Texans love—plate-size chicken-fried steaks with peppery cream gravy, cabrito (roasted or pit-cooked baby goat), and larger-than-life T-bone steaks. When the roads lead you to charming small Hill Country towns, you'll find an amazing array of restaurants—bakeries that beckon patrons bright and early each morning with the enticing aromas of freshly baked pastries and breads; decades-old diners, little changed by the passing of time, where locals gather for the daily specials; quaint tearooms, many of them doubling as gift and antique shops and serving delicious dainty salads, soups, and quiches that we girls love and the men we drag along tolerate; and fine dining establishments serving upscale foods with their roots in the region's rich heritage, in dining rooms decorated in the Hill Country style of understated, slightly rustic elegance.

Don't be surprised if you come upon a festival paying homage to a local crop, specialty food, or regional wines, or just the season! Maybe you'd like to brush up those culinary skills while enjoying the great Hill Country outdoors. That can happen—some of our Hill Country B&Bs, inns, and resort ranches along those winding roads offer cooking classes in conjunction with your stay.

The folks who call the Hill Country home are a special attraction themselves. Those from families who have been in the Hill Country for generations have roots that go back to the Texas war for independence against Mexico. But the region also has a strong siren's song that seems to be heard by the creative, entrepreneurial spirits who have burned out on corporate life in big cities. They come from all over to call the Hill Country home. They let their hair down. They let their hair grow. They trade in three-piece suits and spike heels for jeans and boots. They buy land and establish farms and vineyards, build wineries, open bakeries, restaurants, cooking schools, wine bars—you name it. The lure of the Hill Country is so strong that newcomers create niches for making a living in order to live here. The typical escapee to the Hill Country is well-educated and usually a bit on the eccentric side. But one thing they have in common is that they're a really friendly lot, the kind of folks who never met a stranger. So be prepared for some great conversations as you visit.

Welcome to the rich bounty of the Texas Hill Country, where there's something for every visitor. If this is your first visit, we bet it won't be your last one!

ROASTED POBLANO CAESAR SALAD WITH CORN BREAD CROUTONS

This fabulous salad is one of the most popular items on the Grotto menu. To break up the preparation, make the dressing one day and the croutons the next day—then simply put the salad together. The salad is hearty enough to serve as a luncheon entrée with some hot bread and a nice glass of chilled sauvignon blanc or viognier. The leftover dressing will keep in the refrigerator for about a week.

Yield: 6 to 8 servings

12 cups chopped romaine lettuce
Grated Cotija cheese
Chopped cilantro
Freshly squeezed lemon juice
Freshly cracked black pepper
Mexican chile-limón powder (equal parts lemon pepper and chili powder)
Corn Bread Croutons (see recipe below)

Dressing

2 tablespoons fresh lemon juice, plus a little more for finishing the salad
3 teaspoons anchovy paste
1½ teaspoons Worcestershire sauce
2 teaspoons minced garlic
1 teaspoon black pepper
1 tablespoon minced cilantro
½ cup chopped roasted poblano chilies
1½ teaspoons green curry paste (available in Asian markets)
1 quart real mayonnaise

Corn Bread Croutons

½ cup polenta or coarse cornmeal
1 cup buttermilk
1 tablespoon Texas wildflower honey
1 tablespoon melted butter
2 eggs
⅔ cup all-purpose flour
2 tablespoons sugar
2 tablespoons brown sugar
2¼ teaspoons baking powder
⅛ teaspoon baking soda
½ teaspoon kosher salt
10 slices applewood-smoked bacon, cooked until crisp, drained, and cut into tiny bits

(continued)

The Grotto Grill and Coffee Bar

Bandera

The Grotto Grill and Coffee Bar is one of Bandera's most pleasant surprises. Within the city's concentration of cowboy culture owner-chef Jason Boyd has created an inviting little space, which he describes as urban-retro-modern, and filled it with local art and tables that he designed and made himself. In fact, he did everything inside the restaurant and even built the kitchen. ❖ Jason is a self-taught culinary master who says he acquired his passion for food from his mother, grandmother, and aunts, all marvelous cooks who had vegetable gardens. California chefs Thomas Keller of the French Laundry in Yountville and Alice Waters of Chez Panisse in Berkeley are his main culinary influences. He loves to make people happy with food, noting how much of our lives we spend together around a table sharing food and conversation. With the Grotto Grill, he wanted to create simple food in a casual atmosphere, as though you were a guest in his home. ❖ Jason has definitely succeeded in his goal. The food, made with the freshest ingredients available, is simple in nature but complex in taste. Menu items feature local, organic, seasonal foods, and eclectic wines are served as well as organic, fair-trade, shade-grown, bird-friendly coffees. Nine-five percent of everything available at the restaurant is made on the premises. ❖ The Grotto serves lunch from Tuesday through Saturday, brunch on Sundays, and dinner on Fridays and Saturdays. The lunch menu is loaded with all sort of tempting goodies from focaccia bread as well as hand-made pizzas with a fabulous crust and myriad toppings like Caribbean Afternoon Delight, The Big Easy, and Sergeant Pepper. There are pasta dishes, soups, meal-size salads, and a sandwich for every whim. The brunch menu offers eggs combined with everything but the kitchen sink and even a good ol' fried egg sandwich. ❖ The dinner menus change weekly and feature dishes that would cost you a small fortune in the big city—Duck Soup with Wild Mushrooms, Chilean Sea Bass with Parsnips, Spinach & Saffron-Vanilla Sauce, Beef Cheeks on Watercress & Quinoa with Horseradish Cream & Roasted Beet Puree, to name just a few, and all at a mere pittance compared with the big-city price.

907 13th Street • (830) 796-9555

Make the dressing. Combine all ingredients in blender except the mayonnaise. Puree until smooth. Turn mixture out into a mixing bowl and whisk in the mayonnaise to make a smooth dressing. Adjust seasonings to taste. The dressing should have a fairly pronounced lemon and garlic balance, with just a hint of cilantro and a slight sting from the poblano chilies. Refrigerate until ready to use.

Make the Corn Bread Croutons. Preheat oven to 350 degrees. Grease a large baking sheet and set aside. Stir the polenta and buttermilk together in a large bowl. Whisk the honey and butter together, then add to the polenta. Add the eggs and beat to mix well; set aside.

In bowl of electric mixer combine the flour, sugar, brown sugar, baking powder, baking soda, and kosher salt. Add the buttermilk mixture and beat, starting at low speed and increasing speed as dry ingredients are moistened; mix until well blended. Turn mixture out into the baking sheet; form a very thin layer (about ⅝ inch thick) and scatter the chopped bacon over the entire surface, pushing it into the polenta mixture. Bake in preheated oven for about 30 minutes, or until firm and just turning brown. Cool. Cut into crouton-size squares and turn out onto a clean baking sheet. Reduce oven heat to 300 degrees and bake for 20 to 30 minutes, or until dry and crisp. (Leave the oven door open so the moisture can escape and the croutons can dry out completely.) As the croutons brown, watch carefully and take care not overbrown them. Cool on a wire rack and store in an airtight container until ready to use.

To assemble the salad, toss the chopped romaine with the desired amount of dressing and Cotija cheese. Place a serving of the greens on each plate and garnish with the cilantro, lemon juice, black pepper, and Mexican chile-limón powder. Sprinkle with some of the Corn Bread Croutons and serve.

(The Grotto Grill and Coffee Bar)

Uncle John's Turkey Sausage

Bandera

"**U**ncle John" Pennell, an avid deer hunter for most of his life, was always making venison sausage and constantly trying new recipes, using different critters and various combinations of spices. ❈ One year a friend provided him with some meat from show turkeys, so John made a huge batch of turkey sausage, using some of his secret seasonings and combining the turkey meat with pork to add fat. The sausage was a huge hit with everyone who tasted it. ❈ Then John was encouraged to try making sausage with quail meat from the Diamond H Ranch in Bandera. He marketed the quail sausage for a number of years to a devoted following. But when the Diamond H ceased operations in 2009, John went back to making his sausage from turkey. Most folks seem to like it even better! ❈ Uncle John's Turkey Sausage is not yet a household name, but many area chefs have discovered it and put it on their menus. They love the great flavor pairings they can create with the sausage, and it is proudly served at restaurants in Bandera, Fredericksburg, Kerrville, and numerous other locations.

You can also find Uncle John's Turkey Sausage at an increasing number of places in the Hill Country, including Bandera Wine & Spirits; the Chisholm Trail Winery and Torre di Pietra Winery in Fredericksburg; and Love Creek Orchards and the Oldtimer in Medina.

(830) 460-7655

Alamosa Wine Cellars

Bend

Winemaker Jim Johnson, who owns Alamosa Wine Cellars with his wife, Karen, has often been referred to as "the future of Texas wine." Anyone who has tasted his wines will surely agree. Jim has bucked the tide of tradition in Texas wines, always maintaining that Texas wineries should be growing those varietals that are adaptable to our hot climate. So he didn't plant cabernet or chardonnay or merlot, but rather Mediterranean varietals like tempranillo (from which he makes his award-wining El Guapo), sangiovese, syrah, chenin blanc (from which he crafts Jacque Lapin), and viognier (Alamosa's Viognier is one of the best in the state). After watching the success of Jim's wines, many other Texas wineries are slowly coming around to his way of thinking. He's uncovered the real niche for Texas wines with varietals from Italy's Tuscany, Spain's Rioja, and the Rhone Valley in France. ❋ After working many years in the "oil patch" in Houston, he could no longer ignore the lure of the different drummer's tune. In his mid-forties he packed up everything he owned and headed to California, where he graduated from the University of California's Davis campus with a degree in enology and viticulture. He hung around California for a few years, working at some pretty impressive wineries—such as Heitz Cellars and St. Francis—but Texas was home. With the wine industry growing here, he decided it was time to return. ❋ After a stint with an upscale retail wine shop in Houston, he moved to the Hill Country as the opening winemaker for a new winery. But he longed for his own winery, and it was a concept that Karen wholeheartedly supported. He used to worry that he'd be sixty by the time they got grapes planted. She told him he'd be sixty anyway, so why not have a winery, too! ❋ Alamosa Wine Cellars is one of

the most remote of the Hill Country wineries, located in the sparsely populated, scenic rolling hills west of Bend off Highway 580. The winery was built in 1999, just in time for the harvest that year. It was constructed only as a production facility because Texas law prohibited the retail sale of wine in "dry" jurisdictions at the time. ❋ After the law was changed in 2001, Jim and Karen set up a tasting bar for weekend tastings. In 2004 they expanded the winery to include more barrel storage and a spacious tasting room with a fireplace made of San Saba limestone, which is very inviting in the winter months. The retail space is furnished with antique fixtures from Karen's parents' mercantile store, its venerable counter serving as the tasting bar. The new facility has a covered veranda, the perfect place for picnics or just enjoying a glass of wine while gazing out over the vineyard and hills. The Johnsons run the tasting room themselves, and you'll feel welcome from the minute you turn onto the lavender-lined road that leads to the winery. Jim loves to talk wine!

677 CR 430 • (325) 628-3313 • www.alamosawinecellars.com

HERB-BAKED GOAT FETA CHEESE WITH GARLIC, KALAMATA OLIVES, AND ALMONDS

Yield: 4 to 6 servings as an appetizer

8 ounces CKC Farms Goat Feta cheese cubes
16 pitted kalamata olives
4 large garlic cloves, cut into thin slices
16 blanched whole almonds
1 cup olive oil
1 teaspoon minced fresh rosemary
1 tablespoon minced fresh parsley
1 teaspoon crushed red pepper flakes
1¼ teaspoons freshly ground black pepper
Salt to taste
French baguette, cut into 1-inch slices

Preheat oven to 425 degrees. Place the cheese in an oven-proof 8- to 10-inch oval au gratin dish. Scatter the olives, garlic, and almonds around the dish. Combine the olive oil, rosemary, parsley, red pepper, black pepper, and salt in a bowl; whisk to blend well and flavor the oil. Pour the oil mixture evenly over the cheese. Place dish in preheated oven and bake for about 10 minutes, or until the cheese is very lightly browned and the oil is bubbling. Let cool for about 6 minutes before serving. Serve with a basket of baguette slices for dipping and spreading.

(Terry Thompson-Anderson)

CKC Farms

Blanco

CKC Farms is one of the Hill Country's newest producers of fine-quality goat cheeses, all made by Chrissy Omo, who first learned the art of making goat cheese when she was barely fourteen and living with her family in Germany and Italy. ❧ Being an adventurous eater, she tried all the local cheeses but developed a great interest in the many styles of goat cheese she discovered. In Bologna, Italy, she would make prolonged visits to goat cheese producers, studying their various methods and asking hundreds of questions. The cheese makers were fascinated that one so young could have such a grand interest in making cheese, so she was welcomed by the masters and learned from them. ❧ When the family returned home to Blanco, with the encouragement of her mother, Adriana, Chrissy began to build her herd of Alpine, La Mancha, and Saaman goats while she was still in high school. Upon graduating, she began to market her cheeses to local farmer's markets. ❧ The cheeses are now available in many upscale markets in Austin and throughout the Hill Country, including McCall Creek Farms Market. Upscale restaurants clamor for allotments of her limited production. CKC Farms produces plain chèvre, a blue-veined goat cheese, goat feta cheese, and several flavored and seasonal blends. ❧ Hill Country cheese lovers eagerly await each new variety that Chrissy produces. In the spring of 2008 she released a sensational Romano cheese, which she perfected on a recent trip to Italy.

17505 FM 32 •
(830) 833-5669

MAMA'S APPLE TURNOVERS

Yield: About 20 turnovers

Apple Filling

10 cups thinly sliced peeled and cored apples
(5 cups Granny Smith and 5 cups Fuji)

½ cup sugar

½ cup firmly packed light brown sugar

½ teaspoon salt

1 teaspoon ground cinnamon

Pinch of freshly grated nutmeg

¾ cup water

2 tablespoons butter

Pastry

3 cups all-purpose flour, sifted

1 teaspoon salt

2 teaspoons baking powder

1 teaspoon sugar

1 cup shortening

6 tablespoons ice water

Glaze

2 cups powdered sugar

⅓ cup hot apple cider

¼ teaspoon vanilla extract

Cinnamon-sugar in shaker can

Make the filling. Combine all ingredients in a heavy 8-quart pot over medium heat. Simmer, stirring occasionally, for about 30 minutes. Taste for sweetness, adding additional sugars if needed. Leave the mixture a little juicy and the apples just a bit firm. Refrigerate overnight.

Preheat oven to 425 degrees. Spray baking sheets with cooking spray. Make the pastry. In a large bowl combine the flour, salt, baking powder, and sugar, blemding well with a fork. Cut in the shortening until the mixture resembles coarse meal. Sprinkle in the ice water a little at a time, using a fork to blend gently just until the dough sticks together and forms a ball. Wrap the dough in plastic wrap and refrigerate for 30 minutes. Divide the chilled dough into small balls about the size of an egg. On a lightly floured work surface, roll each dough portion into a circle about ⅛ inch thick. Place ¼ to ½ cup of the filling in the middle of each dough circle, then fold dough over into a half-moon shape. Crimp the open edge. Place turnovers on the prepared baking sheets and bake in preheated oven until golden brown, about 20 minutes.

While turnovers are baking, make the glaze. Combine the powdered sugar, apple cider, and vanilla, blending well. The mixture will be very thin.

Remove turnovers from oven and cool slightly on wire racks. Brush the glaze on the cooled turnovers with a pastry brush and then sprinkle with the cinnamon-sugar mixture.

(The Deutsch Apple Bakery)

The Deutsch Apple Bakery

Blanco

Located in a bright, apple-red building just a few blocks east of the town square, this bakery is one of the Hill Country's hidden treasures. ❀ The bakery was opened in 1976 by Gene and LaVada Triesch, who wanted to market apples from their local orchard. It was purchased in 2004 by Ron and Connie Endress. Connie used to stop at the bakery to buy pies, and she was always impressed by the bakery's commitment to home-baked freshness. When she noticed a "For Sale" sign in front of the bakery, she knew it was a venture with her name written all over it! ❀ To support the bakery's growing popularity, Ron and Connie expanded the kitchen and remodeled the dining area, creating a cozy, inviting space in which to enjoy a cup of good coffee and a slice of pie or a muffin. While still specializing in baked goods made with apples, such as those sinful apple pies and the original Apple Pecan Cake (food of the gods!), the bakery now offers goodies based on Endress family recipes. These include the Ruthie's Cookies line and popular seasonal favorites made from local fruits and berries—Blackberry Cobbler, Peach Cobbler, Pumpkin Pie, and homemade Peach Ice Cream.

Locally produced jams, jellies, and sauces are also sold here. Most of the apples used at the bakery come from the Hohenroth Farm in Blanco, owned by the Triesches and now leased by Ron and Connie. Dating back to 1887, the farm now also raises peaches and plums. ❀ If you just can't fit a stop at the Deutsch Apple Bakery into your Hill Country itinerary, the bakery ships its freshly baked goodies nationwide through its Web site.

602 Chandler Street (Loop 163)
• (830) 833-2882 •
www.thedeutschapple.com

4th Street Market

Blanco

First off, the market isn't on 4th Street—although it started out there in an old meat market where owner Lester Coldewey learned the craft of butchering. He eventually bought the meat market and, in due time, relocated it in a new, larger building across the river on Chandler Street (the continuation of 4th Street). He also moved the huge old meat scale from the original market. Lester and his wife, Carol, say that over the years customers would bring their children into the market and ask the butcher to weigh them on that big scale to "see how they're growing." ❧ The new market is a squeaky-clean, inviting space with a large meat case loaded with hand-cut roasts, steaks, chops, meat ground fresh daily, and tons of the sausage for which Lester is known far and wide, made from his secret formula. It's some of the best sausage you'll find anywhere! He makes some dynamite jerky and dry-cured sausages as well. The hunting season is a busy time as Lester processes game. ❧ Since the market doesn't ship, you'll just have to visit Blanco to get some of that good sausage!

612 Chandler Street • (830) 833-1194

Heron's Nest Herb Farm

Blanco

Fred and Melanie Van Aken are special folks who've created a special place from a rocky piece of land they purchased in 1994 near Blanco. The couple met in Austin, where Fred was an industrial engineering technician for a pharmaceutical firm and Melanie was a midwife with a wide knowledge of medicinal herbs gained from aiding in the birth of some 1,100 babies. They discovered a mutual interest in gardening and began to experiment with some commonly used herbs at their Austin home. Most of the herbs that interested them, such as lavender and echinacea, thrived. ❦

After building their house on the Blanco property in 1997, they began the switch from home gardening to small-production organic farming. Although beautiful, the land had no topsoil, so they "built" it, beginning with many truckloads of granite sand that they used to break up and condition the hard caliche soil. They wanted to establish the farm as a place where people could come and learn how to be gentle consumers, living in a manner that does not harm the earth or their bodies. ❦ Today Heron's Nest Herb Farm has herb-lined trails leading to a central picnic spot. A rainwater collection system waters the herb beds with a solar-powered drip irrigation system. The Van Akens make a wide variety of handmade, all-natural products from herbs grown on the farm or sourced from suppliers of all-natural ingredients. Among the best-selling items is a household cleaner/degreaser made from lavender, fir, and orange. Other products include air freshener and other sprays, shower gel, lotions, salves, essential oils, soaps, sachets, candles, and pet shampoo. Fred has even developed a "compost tea" fertilizer that is very popular with other area farmers. ❦ Melanie teaches classes at the farm on how to make various herbal products; in conjunction with the Blanco

Lavender Festival, the farm offers cooking classes in the many culinary uses of lavender. The farm also schedules seminars featuring noted herbalists and cookbook authors, and classes on soil conditioning are planned. Information about classes and other activities can be found on the Web site. ❦ Heron's Nest Herb Farm is open to the public on selected Saturdays from April through July (see Web site for schedule).

1673 River Bend Drive • (830) 833-2627 • www.heronsnestherbfarm.com

APPLE COOKIES WITH LAVENDER GLAZE

These delicious spiced apple cookies are made with chopped nuts and raisins and topped with a glaze flavored with lavender buds.

Yield: About 3 dozen cookies

2 cups all-purpose flour, stirred before measuring
1 teaspoon baking soda
½ teaspoon salt
1 teaspoon ground cinnamon
½ teaspoon ground cloves
¼ teaspoon ground nutmeg
½ cup shortening
1¼ cups brown sugar, firmly packed
1 egg
1 cup chopped walnuts or pecans
1 cup finely chopped apple, unpeeled
1 cup raisins, chopped
¼ cup milk or juice

Lavender Glaze

1 tablespoon butter
3 tablespoons milk, or substitute half and half
2 tablespoons lavender buds
1½ cups powdered sugar, sifted
½ teaspoon vanilla extract
Dash of salt

Make the Lavender Glaze first. Combine butter, milk, and lavender buds in a small saucepan. Heat until almost boiling, turn off, cover, and let sit for about 2 hours. Strain off lavender buds. In a small mixing bowl, sift powdered sugar and salt; add vanilla. Add to the cooled milk mixture. Stir until all is dissolved. Set aside.

To make the cookies, preheat oven to 375 degrees; grease two baking sheets and set aside. Sift flour with baking soda, salt, cinnamon, cloves, and nutmeg. In a large mixing bowl, cream shortening and brown sugar; beat in egg until well blended. Stir in half of the flour and spice mixture, then stir in walnuts, apple, and raisins. Blend in milk, then the remaining flour mixture.

Drop dough by rounded tablespoons onto prepared baking sheets, spacing cookies about 2 inches apart. Bake in preheated oven for 12 to 15 minutes, or until done. While cookies are still hot, spread with a thin layer of Lavender Glaze.

(Melanie Van Aken)

Hill Country Lavender

Blanco

In 1999 photographer Robb Kendrick, on assignment in southern France, was taken with the beauty of the rows of bushy purple lavender. After talking to the lavender farmers in the area, he learned that the soil and climate were comparable to those of the Texas Hill Country. He returned home and began to experiment with planting different varieties of lavender on his acreage near Blanco. ❧ With advice from the French farmers, he and his wife, Jeannie Ralston, planted two acres of lavender in 1999. The crop grew fabulously, and the next year they planted more plants. By the spring of 2002, a total of 6,000 plants bloomed during the blooming season from mid-May into early July. In 2004 Rob and Jeannie installed a commercial-size distilling facility to create lavender oil from buds and stems. They also began to host local seminars on growing lavender, paving the way for a new Hill Country industry. ❧ Hill Country Lavender was purchased by Tasha Brieger in 2006 when its original owners decided to move to Mexico. Tasha, formerly Robb's photo assistant, had worked on the farm for six years when she was made its manager. ❧ With the change in ownership, the farm was moved to its current location on a two-acre hilltop behind McCall Creek Farms Market just over three miles north of Blanco. From the end of May to the end of June, you can cut your own lavender at the farm. The store, snuggled into the lavender field, offers more than seventy lavender-related products for sale. Hill Country Lavender products are also available year round at Brieger Pottery, owned by Tasha's parents and located on the town square in Blanco.

4524 N. U.S. 281 • (830) 833-2294 • www.hillcountrylavender.com

McCall Creek Farms Market

Blanco

After several years of growing cut flowers for the wholesale trade, Mark and Cathie Itz, both raised in agriculture in the Texas Hill Country, decided to switch to produce. In 2000 they opened the McCall Creek Farms Market, embracing their dream of operating a real country market. Located several miles north of Blanco, the market has a devoted following of folks who drive for miles to purchase the farm's quality

fruits and vegetables as well as produce from surrounding farms, such as Amador Farms lettuces. ❀ McCall Creek Farms Market sells seasonal produce and peaches grown on the farm as well as blackberries grown at Itz Gardens, the farm of Mark's father, Melvin Itz. The kitchen inside the market is where Cathie makes an eagerly awaited daily supply of fresh breads, pies, cookies, four varieties of homemade ice cream, and peach cobbler in season. And her whole-grain and nut granola is truly the best you'll ever eat. The Itzes' youngest son, Skylar, has a flock of free-range chickens and sells his hens' eggs at the market. Patrons can also buy salsas made by other family members, locally produced jams and pickles, pecans, CKC Farms cheeses, and Loncito's Grass-Fed Lamb products. ❀ McCall Creek Farms consists of land that Cathie inherited from her father's ranch, Conn Ranch, which originated in 1874 with a State of Texas land grant. Through the generations, the ranch grew in size and was inducted into the Texas Family Land Heritage Registry in 1998. Mark and Cathie live in the home that her great-grandfather, J. Lee Conn, built on the ranch about 1905. ❀ Mark sums up the life of a farmer succinctly: "We're not billionaires here—just real farmers trying to keep the land in the family."

4524 N. U.S. 281 • (830) 833-0442 • www.mccallcreekfarms.com

ICEBERG WEDGE WITH BLUE CHEESE DRESSING

Don't discount iceberg lettuce as an inferior salad green. Nothing makes a more impressive salad than a tall quarter wedge of iceberg lettuce draped with a great blue cheese dressing. Fresh-from-the-ground iceberg has a marvelous flavor all its own and the ultimate crispness to be found in a salad green.

Yield: 4 servings

1 head iceberg lettuce, cored, sliced into 4 wedges, and well chilled

6 slices applewood-smoked bacon, cooked until crisp and chopped into fine dice

4 ounces (1 cup) crumbled blue cheese

Blue Cheese Dressing

½ cup real mayonnaise

½ cup sour cream

1 teaspoon celery salt

2 teaspoons freshly squeezed lime juice

2 teaspoons Dijon-style mustard

1 tablespoon minced onion

1 tablespoon minced fresh Mexican mint marigold, or substitute minced fresh tarragon

1 teaspoon Tabasco

3 ounces blue cheese

Make the Blue Cheese Dressing. In work bowl of food processor, combine all ingredients except the blue cheese. Process until smooth and well blended. Add the blue cheese and pulse just until the cheese is blended; leave some small lumps. Refrigerate until ready to serve.

To serve the salad, cut a small slice off the bottom of each wedge so that it will stand flat on the plate. Place one wedge on each serving plate. Drape a portion of the dressing across each wedge, allowing the dressing to pool slightly on the plates, then scatter some of the bacon and crumbled blue cheese over the lettuce and around the plate.

(Terry Thompson-Anderson)

BREWHOUSE BROWN ALE AND CHEDDAR CHEESE SOUP WITH GARLIC FLOATERS

Real Ale Brewing Company's Brewhouse Brown Ale is a caramel-brown ale made from four kinds of malts and American hops. It's the perfect taste complement in this hearty, delicious soup.

Yield: 10 to 12 servings

¼ cup canola oil

2 medium onions, chopped

3 carrots, peeled and sliced thin

3 celery stalks, chopped

4 baking potatoes (about 2½ pounds), peeled and cut into ½-inch dice

4 jalapeños, seeds and veins removed, minced

⅓ cup whole-grain mustard

1 bottle (12-ounce) Real Ale Brewing Company Brewhouse Brown Ale, or substitute another European-style dark bock beer

2 quarts chicken stock

3½ cups (14 ounces) shredded sharp Cheddar cheese

Salt to taste

1 cup whipping cream

Garlic Floaters

4 cups ½-inch French bread cubes

¾ pound (3 sticks) unsalted butter, melted (1½ cups)

¼ teaspoon salt

2 heaping teaspoons granulated garlic

Make the Garlic Floaters. Preheat oven to 300 degrees. Place the bread cubes in a medium bowl. In a separate bowl, whisk together the melted butter, salt, and granulated garlic. Pour the mixture evenly over the bread cubes and toss quickly, coating well. Turn the bread cubes out in a single layer onto a baking sheet. Bake in preheated oven until cubes are lightly brown and very crisp. Turn out onto a wire rack and cool completely. To store the Garlic Floaters for up to three days, transfer to a container with a tight-fitting lid.

To make the soup, heat the canola oil in a heavy, 6-quart soup pot over medium heat. Add the onions, carrots, celery, and potatoes. Cook, stirring occasionally, until onions are wilted and transparent, about 8 minutes. Add the jalapeños, mustard, beer, and chicken stock. Stir to blend well, cover, and simmer for 30 minutes, or until potatoes are very soft and have started to break apart. Puree the soup in batches in a blender or food processor until completely smooth. Return to a clean pot over medium heat and stir in the cheese. Cook just until the cheese has melted. Add salt to taste. Do not allow the soup to boil once the cheese has been added. Stir in the whipping cream and cook just to heat the cream through. Serve hot, garnished with Garlic Floaters.

(Terry Thompson-Anderson)

Real Ale Brewing Company

Blanco

The Real Ale Brewing Company makes some serious beer in an artisan environment that fosters unique beers. It's all about passion and commitment to the quality of the product. ❋ The company's president, Brad Farbstein, started brewing beer in 1987 as a hobby while he attended the University of Texas at Austin. He loved the whole process of creating beers with great flavor. After working as a marketing director for a craft brewery in Houston, he was able to buy Real Ale in 1998. At that time, it was a tiny brewing company located in the basement of an antique shop on Blanco's town square and produced 600 barrels of beer. In 2007 Brad built a new, state-of-the-art brewery that could produce 30,000 barrels. ❋ The entire staff is totally involved with the products from beginning to end, following an exacting process supervised by plant manager Todd Ehlers and head brewer Tim Schwarz. ❋ With its 60-barrel stainless steel brewing system, the company makes a fabulous lineup of beers using 100 percent malted grain, domestic and imported hops, and crystal-clear water from the Blanco River. Several beers—Rio Blanco Pale Ale, Full Moon Pale Rye Ale, Brewhouse Brown Ale, and Fireman's #4 Blond Ale—are available year round at most bars, restaurants, and grocers in Austin, Houston, Dallas, Fort Worth, San Antonio, and around the Hill Country. The company also produces some unique seasonal brews: Sisyphus Barleywine Ale, Phoenixx Double Extra Special Bitter, Devil's Backbone, Lost Gold IPA, Roggenbier, and Shade-Grown Coffee Porter. ❋ The brewery is open to the public on Friday afternoons, when tours are available.

231 San Saba Court
(830) 833-2534
www.realalebrew-ing.com

Texas Lavender Hills Farm & Market

Blanco

Third-generation Blanco landowners Jill and Doak Hunter founded Texas Lavender Hills Farm & Market in 2004, envisioning exciting new possibilities for their 27-acre property, part of the 300-acre ranch where the Hunter family had lived for sixty-five years. ❋ Inspired by the lush lavender fields in France's Provence region, they planted their scenic hillside with over 4,500 lavender plants that they are growing with organic methods. The 360-degree view from the crest of the lavender field is spectacular. ❋ In the valley below the lavender fields the Hunters have built a lovely market that is open from Memorial Day weekend through the end of the lavender growing season in late June. The market features lavender plants of many varieties for sale along with local art, lavender personal care products, teas, and much more. ❋ Texas Lavender Hills also has a year-round shop in downtown Blanco at the Vintage Cottage at 508 4th Street.

5110 Kendalia Road • (830) 833-9183/833-4709 • www.texaslavenderhills.com

CHILI CORN CREPES

Chef Weber serves these crepes on a bed of pureed black beans surrounded by grilled squash.

Yield: 4 servings (8 to 10 crepes)

1 cup yellow cornmeal
¼ cup all-purpose flour
1 tablespoon ancho chili powder
1 teaspoon salt
1 large egg
¾ cup whole milk
1 tablespoon melted butter

Chipotle Sauce

2 chipotle chilies in adobo, chopped
1 cup whipping cream
1 teaspoon salt

Lime-Butter Sauce

¾ cup freshly squeezed lime juice
6 ounces (1½ sticks) unsalted butter, cut into
   ½-inch dice

Filling

2 (6-ounce) spiny lobster tails
1 cup lump crabmeat
Butter for sautéing
1 large avocado, peeled and cut into ½-inch dice
1 large ripe mango, peeled and cut into
   ½-inch dice
1 teaspoon salt
1 teaspoon black pepper

Combine the cornmeal, flour, chili powder, and salt in a small mixing bowl. In a separate bowl, combine the egg, milk, and butter. Add to the cornmeal mixture, blending well. Let batter stand for 30 minutes.

Heat a nonstick 6-inch omelet or sauté pan over medium heat until a drop of water placed in the pan dances and evaporates almost immediately. Ladle 2 tablespoons of the crepe batter into the center of the hot pan and swirl the pan quickly until the batter covers the bottom of the pan. Cook 30 seconds, gently flip the crepe, and cook an additional 30 seconds on the other side. Repeat until all batter is used.

(continued)

Uptown Blanco Arts & Entertainment Restaurant

Blanco

It's not every day that someone comes along and buys an entire block of a Hill Country town. But that's exactly what happened in Blanco when Renee Benson bought the 300 block of Main Street on the town square. ❖ Renee, daughter of New Orleans Saints owner Tom Benson and a thirty-year resident of the Hill Country, got the idea for an arts and entertainment complex in 1999, when she noticed a group of buildings in various states of disrepair. She began buying and renovating the properties as they became available. ❖ The restaurant, housed in a former general merchandise store, opened in 2006. The renovation was extensive, creating a marvelous space reminiscent of a grand restaurant of the late nineteenth century, when the building was erected. With breakfast, lunch, and dinner menus, the restaurant enhances the experience of patrons attending other Uptown Blanco venues, such as the theater (located in the adjacent building), textile studio, tavern, ballroom, art center, frame shop, and courtyard. ❖ The Uptown Blanco kitchen is under the direction of Chef Robert Webber, who has worked with some of the state's premier chefs, including Tim Keating and Bruce

Auden. Chef Webber's menus are a delight—an eclectic romp through the various cuisines that have influenced his style, such as Cajun, Southwestern, and East Asian. Many of his dishes feature products from local farms.

317 Main Street •
(830) 833-0738 •
www.uptownblanco.
com

Make the Chipotle Sauce. Combine all ingredients in a small saucepan and cook over medium heat until cream is thickened enough to coat the back of a spoon. Strain through a fine strainer and set aside; keep warm.

Make the Lime-Butter Sauce. In a small saucepan over medium-high heat, reduce the lime juice to about 2 tablespoons. Reduce heat to low and whisk in the butter, one piece at a time, until all has been added (the sauce will be thin). Remove from heat and set aside; keep warm, but not hot, or the sauce will separate.

To make the filling, split the lobster tails in half and remove the meat; dice and set aside. Pick through the crabmeat to remove any shell fragments; set aside.

To assemble the crepes, place crepes on a baking sheet in a warm oven. Sauté the lobster meat in a small amount of butter. Add the crabmeat, mango, and avocado; season with salt and pepper. Cook just until warm. Remove from heat and stir in the Lime-Butter Sauce. Divide the lobster mixture into 8 equal portions and place in the center of the crepes; roll crepes up and place on serving plates. Drizzle Chipotle Cream over the crepes and serve warm.

(Uptown Blanco Arts & Entertainment Restaurant)

MUSHROOMS AND AURELIA'S CHORIZO IN TRIPLE SHERRY CREAM SAUCE

This recipe makes a pretty sensational party dish.

Yield: 4 to 6 servings as a finger food

3 tablespoons pure olive oil

2 large shallots

3 large garlic cloves, minced

5 ounces small white button mushrooms, halved

3 ounces shiitake mushrooms, sliced

¼ teaspoon crushed red pepper flakes

2 cups medium-dry sherry (amontillado), divided

3 thyme sprigs

1 link Aurelia's Chorizo (about 2½ ounces), sliced very thin

2 cups whipping cream

1 Knorr chicken bouillon cube

Salt and freshly ground black pepper to taste

½ cup cream sherry

French baguette slices, slightly toasted

Heat olive oil in a heavy 10-inch skillet over medium heat. When oil is hot, add the shallots and garlic; sauté until shallots are wilted and transparent, about 5 minutes. Do not allow garlic to brown. Stir in the mushrooms and crushed red pepper. Cook, stirring often, until mushroom liquid has evaporated and mushrooms are limp, about 6 to 7 minutes. Add ⅔ cup of the dry sherry and the thyme. Stir in the chorizo. Cook over medium-high heat, stirring often, until sherry is reduced to a mere glaze, about 5 minutes. Repeat two more times, using ⅔ cup of sherry and reducing to a glaze each time. Stir in the whipping cream, scraping up any browned bits from the bottom and sides of the pan. Add the bouillon cube and stir to dissolve. Season to taste with salt and freshly ground black pepper. Cook over medium heat, stirring often, until cream is thickened, about 6 minutes. Add the cream sherry and stir to blend well. Adjust seasoning. Remove from heat and discard the thyme sprigs. Transfer to a shallow serving dish and serve hot with toasted baguette slices.

(Terry Thompson-Anderson)

Aurelia's Chorizo

Boerne

Aurelia's Chorizo was born from Leslie Horne's addiction to authentic Spanish chorizo, which a friend smuggled from Spain twenty years ago. It was a classic case of love at first bite—a bite that started Leslie, a former homemaker, on an odyssey into the world of making this dry-cured sausage. ✳ Leslie researched recipes and began to make the chorizo for friends. But it was hard to cold-smoke the sausage properly at home. So a determined Leslie went to Spain in 2000, touring chorizo plants and learning how to make the sausage commercially. When she finally found a processor with a kindred adventurous spirit, they worked together to build the smoking and drying facility according to USDA specs. ✳ With the formula finally perfected, Leslie began producing Aurelia's Chorizo in 2006. The sausage is made to very high standards. Full cuts of pork and beef chuck are used with absolutely no fillers or trimmings. Garlic is freshly ground for each batch of the sausage, and Leslie imports pimentón—the smoked paprika that is so important to the sausage's deep, rich flavor—from the La Vera region of Spain, known for its fine paprika. The smoking, aging, and drying process takes an entire month, and the result is a really fabulous, very authentic Spanish chorizo. Aurelia's Chorizo is available at Central Market stores and at Spec's Liquor Stores in Austin.

(830) 446-1325 • www.aureliaschorizo.com

Bear Moon Bakery

Boerne

The stated mission of Paula Hayward's community bakery is "to nourish the Community Heart." Mission accomplished, Paula. ✼ Step inside the bakery in the wee hours of the morning and your senses are immediately aroused by the aromas of wonderful things baking in the massive ovens in back. By 7 a.m. the pastry cases are loaded with pastries, cookies, cakes, scones, muffins, teacakes, pies of every description, and such rare finds as Kugelhopf, a light yeast cake of Austrian derivation filled with raisins, candied fruits, and nuts and generously dusted with powdered sugar. ✼ The place is bustling with sleepy-eyed folks picking up preordered trays of pastries, cookies, or cakes. Others are seated at tables overlooking Boerne's busy Main Street, enjoying their coffee (consistently voted as "the best coffee in Boerne") and breakfast pastries or items from the breakfast buffet. A favorite is the Jalapeño Hog, a smoked sausage in pastry with cheese and sliced jalapeños—the bakery sells over 10,000 per week! ✼ By mid-morning, when the focus shifts to lunch preparation, Bear Moon once again fills up with folks hungry for the from-scratch soups, an array of salads, and sandwiches served on homemade breads. ✼ Housed in a building dating from the early 1900s, the bakery opened in the 1960s under the name Hill Country Bakery. In the 1980s Boerne's longtime mayor, Patrick Heath, was the baker. Paula—a photographer, artist, and chef who found food to be the niche she was seeking—purchased and renamed the bakery in 1995. ✼ A strong believer in community involvement and sustainability, Paula cuts energy con-

sumption with a mix of lighting and timers and recycles plastics, metal, glass, and newsprint. All produce trimmings and coffee grounds are composted. The bakery grows its own herbs and flowers, uses roof runoff water for irrigation, and works with local farmers and growers. ✼ Bear Moon, a favorite among locals, has been named "Best Bakery" for over eleven years. It is a must-visit stop in Boerne and has been featured in *Texas Monthly* and *Texas Highways*, boosting its already excellent reputation.

401 S. Main Street • (830) 816-2327

Cheezy Pigs
1.65

APPLE CAKE

This delicious confection is a special Christmas item at the bakery, but it's easy to prepare at home all year round.

Yield: One 9-inch cake

3 cups all-purpose flour
1 teaspoon baking soda
½ teaspoon salt
2 cups sugar
1 cup vegetable oil
3 large eggs
1 teaspoon vanilla
1 cup chopped pecans
4 medium Granny Smith apples, peeled, cored, and chopped (3 cups)

Glaze

1 cup brown sugar
⅓ cup butter
⅓ cup milk

Preheat the oven to 350 degrees. Butter and flour a 9-inch springform cake pan; set aside. Sift together the flour, baking soda, and salt. In bowl of electric mixer at medium speed, beat together the sugar, vegetable oil, eggs, and vanilla until pale yellow and smooth. Beat in the flour mixture, a third at a time, scraping down the sides of the bowl after each addition. Mix in the pecans and apples. The batter will be very thick. Turn out into the prepared springform pan and bake in preheated oven for 45 to 55 minutes, or until a toothpick inserted in the middle of the cake comes out clean. Cool in pan for 10 minutes, remove sides and bottom of pan, and place cake on cooling rack. Cool completely before glazing.

While the cake is baking, prepare the glaze. In a 2-quart saucepan, bring all ingredients to a boil, reduce heat, and simmer for 5 minutes. Brush or spoon over the cooled cake.

(Bear Moon Bakery)

PANCETTA-WRAPPED PORK TENDERLOIN WITH BASIL-POTATO CRESPELLE STUFFED WITH PORTABELLA MUSHROOMS AND TOMATO–OLIVE OIL SAUCE

Yield: 4 servings

2 trimmed pork tenderloins (about 1½ pounds total), silver skin removed

Salt and freshly ground black pepper

¼ cup thinly sliced basil leaves

1 cup chopped fresh mozzarella

16 thin slices of pancetta or bacon, rolled thin with a rolling pin

Olive oil

Potato Crespelle

2 large baking potatoes, baked until fork tender, cooled, and peeled

3 eggs

1¼ cups milk

2 tablespoons canola oil

Pinch of freshly grated nutmeg

Pinch of ground ginger

½ cup all-purpose flour

1 tablespoon minced basil leaves

Salt and ground white pepper

Olive oil

Portabella Filling

2 large portabella mushrooms, stems removed

2 tablespoons balsamic vinegar

½ cup olive oil

Salt and black pepper to taste

12 pieces pencil asparagus, trimmed and blanched in salt water until crisp-tender

Shredded Parmesan cheese

Tomato–Olive Oil Sauce

4 Roma tomatoes, chopped

1 red bell pepper, chopped

1 tablespoon peeled and crushed garlic

½ teaspoon crushed red pepper flakes

½ cup chopped red onion

1 cup olive oil

¼ cup red wine vinegar

1 rosemary sprig

Salt and freshly ground black pepper to taste

(continued)

Cypress Grille

Boerne

Cypress Grille is entered through the historic Napa Building amid the hustle and bustle of Boerne's busy Main Street. Opened in 2006, the restaurant was the brainchild of Paul Thompson, Tom Stevens, and his wife, Shanda. The trio wanted to serve great food with fabulous wines, all in a comfortable environment. ❉ Cypress Grille definitely has one of the Hill Country's most sophisticated wine lists and a wine room that is not to be missed. And that fabulous real draft root beer is just the ticket to quench a big thirst after shopping Boerne's Main Street. ❉ The restaurant has been a success since its opening, with a menu featuring an eclectic mix of "upscale down-home" offerings and downright upscale treasures. The lunch menu features appetizers that present an array of innovative flavors, a large selection of salads and sandwiches, soups, a daily lunch special, and, of course, housemade desserts. The dinner menu pulls out all the stops, from a lengthy selection of appetizers (including—be still my heart—foie gras with a port reduction!) to steaks and chops, seafood, poultry dishes, and

even more desserts. Cypress Grille has become one of Boerne's most popular dining destinations. ❉ Paul, who owns the business with other family members, spent many years in California, where he learned the craft of butchering and also gained an extensive knowledge of wines. After returning to Texas, he graduated from the Texas Culinary Academy in Austin. Tom, the chef at Cypress Grille, graduated from the Culinary Institute of America in Hyde Park, New York. Shanda runs the "front of the house," greeting customers with a smile that says "Welcome to our table."

170 S. Main Street, Suite 200 • (830) 248-1353
• www.cypressgrilleboerne.com

Make the Tomato–Olive Oil Sauce. Place all ingredients in a heavy 3-quart saucepan over medium-low heat. Slowly simmer for 25 minutes. Remove the rosemary sprig and allow sauce to cool slightly, then pour into blender and puree until smooth. Adjust seasoning as needed. Strain and return to saucepan over very low heat; keep warm.

Butterfly the pork tenderloins from the tip to the tail, slowly unrolling them. Using a meat pounder, lightly pound both tenderloins into rectangles about ¼ inch thick. Season to taste with salt and black pepper. Scatter the basil strips over the meat, then top each with an equal portion of the mozzarella. Starting at the long side, roll up each tenderloin tightly. Lay the pancetta strips out, slightly overlapping, and set the rolled tenderloins on the pancetta strips. Roll up the strips very tight, covering the tenderloin with the pancetta. Season with salt and black pepper and set aside. Preheat oven to 350 degrees.

Make the Portabella Filling. Place the mushrooms on a char grill and brush with vinegar and olive oil. Season with salt and black pepper. Place the asparagus on the grill and season with olive oil, salt, and black pepper. Remove the asparagus as soon as it has grill marks. When the mushrooms are slightly wilted, cut them into thin strips. Set asparagus and mushroom strips aside.

To make the Potato Crespelles, push the potatoes through a ricer or grate into a large bowl. Add remaining ingredients except olive oil and stir batter to blend well. Allow to rest for 30 minutes.

While batter is resting, heat a very thin glaze of olive oil in a heavy 12-inch skillet over medium-high heat. When oil is very hot, sear the tenderloins on all sides until browned. Transfer to a baking dish and finish cooking in preheated oven for about 20 minutes.

While tenderloins are roasting, cook the Potato Crespelles. Heat a 10-inch nonstick pan and oil lightly with the olive oil. Pour in ⅓ cup of the batter and swirl pan to distribute batter evenly. Lightly brown, flip, and cook an additional 20 seconds; remove from pan. Set aside and keep warm; repeat with the remaining batter.

When the tenderloins are cooked, remove from oven and set aside to rest while assembling the Potato Crespelles.

Lay crespelles on work surface and divide the asparagus and mushroom strips among them. Scatter a generous portion of Parmesan cheese over the vegetables, then roll the crespelles up tight. Place 2 crespelles on each serving plate. Slice the tenderloins about ½ inch thick and fan out 3 to 4 slices, slightly overlapping the crespelles on each plate. Top with Tomato–Olive Oil Sauce and serve at once.

(Cypress Grille)

CORN BREAD SALAD

Yield: 8 to 10 servings

2 (10-inch) skillets of baked sweet corn bread

4 ounces (1 cup) shredded Cheddar cheese

⅔ cup finely chopped green onions and tops

1 cup diced fresh tomatoes

6 slices peppered bacon, cooked crisp and crumbled

½ cup Miracle Whip salad dressing, or substitute real mayonnaise

Turn corn bread out of skillets and allow to cool completely, then crumble into fine crumbs in a bowl. Add cheese, green onions, tomatoes, and crumbled bacon. Toss to combine well. Stir in Miracle Whip, blending thoroughly. Chill before serving.

(Hard 8 BBQ)

Hard 8 BBQ

Brady

You've gotta have a lot of moxie to open a barbecue joint in the Texas Hill Country, especially in McCulloch County—the heart of hunting country where folks know how to cook meat. So a barbecue place named for a roll of the dice probably does, indeed, have moxie. ❧ Hard 8 smokes meat the way the region's German settlers did: by allowing the mesquite wood to burn down in a fire pit first and then transferring the glowing coals to the actual cooking pits. With this method, the meats aren't oversmoked. They have a great smoke flavor, but without any hint of resinous sap that you sometimes get from mesquite. ❧ You choose from a carnivore's smorgasbord of brisket, chicken, two kinds of sausage (regular or jalapeño), pork chops, and on weekends honest-to-goodness smoked prime rib, cabrito, and hand-cut, bone-in rib eye. ❧ The array of sides includes potato salad, coleslaw, green salads, pinto beans, and the usual accompaniments of burger pickles, chopped onion, and pickled jalapeños. An unusual and interesting side dish is Hard 8's signature Corn Bread Salad. The house barbecue sauce is a thin, vinegar-based sauce. ❧ Desserts include homemade pies (pecan, buttermilk, apple, coconut, and chocolate), homemade cobblers, and a right-on banana pudding. The restaurant serves a good selection of cold beers as well as frozen margaritas and even wine. ❧ On weekends there's an occasional crawfish boil and live entertainment on the patio. Can't get much more Texan!

2010 S. Bridge Street • (361) 277-3417 • www.hardeightbbq.com

Hairston Creek Farm

Burnet

Gary Rowland was no stranger to growing things when he and his wife, Sarah, purchased their thirty-six-acre farm in 1990. Gary had grown up in the melon fields of East Texas and worked at one of the Texas A&M Agricultural Experiment Stations researching vegetables. ❊ Hairston Creek Farm is a small sustainable family farm that specializes in organic vegetables, herbs, and fruit. The Rowlands' goal is to provide fresh, organic, nutritious local produce in season. Certified organic by the Texas Department of Agriculture since 1993, the farm consists of seventeen acres of mixed vegetables covering the gamut from asparagus to zucchini as well as strawberries and blackberries. A small apiary improves crop pollination, and the Rowlands' two children tend the bees and bottle the honey. Using excess fresh fruit and produce, the Rowlands make jams, jellies, and pesto in their state-certified commercial kitchen, and they raise cage-free chickens and market "yard" eggs. ❊ The Rowland family is totally dedicated to sustainable agriculture. Sarah is very concerned with people's general attitude about food. "They tend to think of food as they do water and electricity—it's just there. They have no idea where it comes from!" Gary adds his thoughts on growing organic: "It just makes sense to grow organically—on a family level, community level, national level, and global level. With every acre of organically grown vegetables and livestock, that's an acre free of chemical fertilizers, insecticides, and herbicides. We just need more of those organic acres to clean up the world!" ❊ Saturday mornings are busy times with the family

selling Hairston Creek Farm products at two separate farmer's markets: the Farmer's Market on the Square in Burnet and also the Sunset Valley Farmer's Market in Austin.

4300 CR 335 •
(512) 756-8380 •
www.hairstoncreekfarm. com

Pan-seared fresh greens make a heavenly winter side dish, especially when made with Hairston Creek Farm's perky, leafy greens. For the best flavor, do use the bacon drippings!

Yield: 4 servings

3 tablespoons bacon drippings

3 large garlic cloves, cut into thin slices

10 cups mixed red chard and green kale, torn into bite-size pieces, with thick mid-ribs removed and discarded

Balsamic Reduction (see recipe below)

6 slices peppered bacon, cooked until crisp, drained well, and chopped into fine dice

Balsamic Reduction

2 cups balsamic vinegar

1 cup firmly packed light brown sugar

Make the Balsamic Reduction. Stir together the balsamic vinegar and brown sugar in a heavy 2-quart saucepan over medium heat. Bring to a boil; reduce heat to a simmer and cook until the liquid is thick and syrupy, about 30 minutes. Cool and transfer to a plastic squeeze bottle. (You won't need all of the reduction for the batch of greens. The remaining reduction may be refrigerated for up to three weeks.)

Heat the bacon drippings in a heavy, deep-sided 12-inch skillet over medium heat. When the fat is hot, add the sliced garlic and greens, tossing to coat with the drippings. Cook, tossing constantly, until the greens begin to wilt. Squeeze some of the Balsamic Reduction onto the greens and continue to cook, tossing, until the greens are wilted but still crisp. Add the chopped bacon and toss to distribute. Serve hot.

(Terry Thompson-Anderson)

Tea-Licious

Burnet

Owner Vicki McLeod opened her first tearoom in Burnet on Valentine's Day of 1996. She and her husband, Sam, had moved to their ranch outside of Burnet after a long tenure in Montgomery, where Vicki had taught high-school English. After eighteen months of cabin fever on the ranch and no luck in finding a teaching position, Vicki had the idea of opening a tearoom. Sam wasn't terribly excited about the venture but went along with it, asking her only that she not lose too much money. After all, he thought, there are only a few thousand residents in the whole town. Little did he know what the future and Vicki's famous Chicken Salad would bring! ❧ That first day the tearoom drew more than a hundred customers, and word spread quickly about the delicious array of sandwiches, specialty salads, quiches, soups, and an awesome loaded baked potato. Desserts range from homemade pies to cheesecakes and individual Chocolate Volcano cakes. ❧ Within two years Vicki had moved into a 5,000-square-foot historic building on the square. With the expansion came a new gift shop–boutique known as Giftique, where Vicki sells many one-of-a-kind, must-have items as well as her Tea-Licious Gourmet Sweet Pickles. Sam eventually retired from his career in medical equipment manufacturing to help run the tearoom! ❧ Billed as "the little tearoom on Burnet's historic square," Tea-Licious is a charming respite from browsing in the community's unique shops. It's a favorite lunch spot of passengers traveling on the *Hill Country Flyer*, an authentic steam locomotive–powered excursion train that the Austin Steam Train Association runs between Cedar Park and Burnet on weekends from January through May and on select weekends in December when Burnet is fully decked out in Christmas finery. There's ample time for shopping and a visit to Tea-Licious before the "All Aboard" signals the return trip with shopping bags full of goodies in tow and memories of a great meal.

216 S. Main Street • (512) 756-4405 • www.tea-licious.com

Camp Verde General Store

Camp Verde

The Camp Verde General Store has a long and colorful history, very much intertwined with the settlement of the Texas frontier. It was established in 1857 by an enterprising man named John Williams, who saw the opportunity to supply provisions to the nearby army outpost on the north bank of Verde Creek. The fort is best known as the headquarters of a corps of imported camels, part of an army experiment in using these animals for military purposes. ❖ In 1858 the store was acquired by Charles Schreiner and his brother-in-law, Casper Real, who obtained a contract to sell beef and wool to the fort. As the operation grew, they offered other items for sale, such as sewing kits, foodstuffs from the surrounding farms, paper and writing materials, blankets, saddle equipment, and eventually liquor (which was forbidden on the fort grounds). A post office operated on the premises in 1858. The store became a gathering place for the local settlers. ❖ Closed during the Civil War, the store reopened in 1887, and a small community began to grow up around it. Numerous owners operated the store after the original structure was destroyed by floodwaters in the early 1900s. ❖ By 2003 the building was in need of repair and restoration when Cynthia Grossman and her husband purchased a ranch above the old site of Camp Verde. Jokingly, their real estate agent asked if they'd like to buy the old store. But Cynthia was intrigued by its history and possibilities and did, indeed, buy it. ❖ Camp Verde General Store has been lovingly restored, and today it operates as a retail store and café. Attractive upscale boutique items, kitchenware, candles, home accessories, jewelry, soaps, food products, artwork, luggage, and oodles of other must-have items are sold in the two-story retail space. ❖ The café offers an eclectic menu of breakfast items, snacks, salads, sandwiches, burgers, pizzas, and desserts, plus a daily "Plate Lunch" listed on the blackboard. The outdoor patio areas offer the perfect venue for a nice breath of fresh air. ❖ Cynthia has great visions for the Camp Verde General Store and intends to add a full-service restaurant in the near future.

285 Camp Verde Road East •
(830) 634-7722 • www.campverdegeneral
store.com

Cherokee Grocery
Cherokee

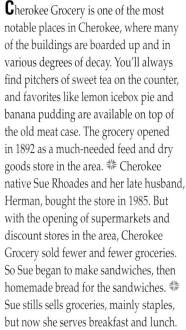

Cherokee Grocery is one of the most notable places in Cherokee, where many of the buildings are boarded up and in various degrees of decay. You'll always find pitchers of sweet tea on the counter, and favorites like lemon icebox pie and banana pudding are available on top of the old meat case. The grocery opened in 1892 as a much-needed feed and dry goods store in the area. ❀ Cherokee native Sue Rhoades and her late husband, Herman, bought the store in 1985. But with the opening of supermarkets and discount stores in the area, Cherokee Grocery sold fewer and fewer groceries. So Sue began to make sandwiches, then homemade bread for the sandwiches. ❀ Sue stills sells groceries, mainly staples, but now she serves breakfast and lunch. She arrives at the store every morning at 6:15 to be ready for the customers who start gathering a little before 7:00. She makes salads, burgers on homemade buns, and anything else she thinks of that folks might like. In addition to the pimento cheese sandwiches, her chicken salad is legendary, and she won't part with the recipe either! And we can't forget those pies, cookies, puddings, and homemade peanut brittle. ❀ Cherokee Grocery is a place where all the locals and transplants gather to catch up on the latest gossip. Also among her regulars are vendors of every sort who travel Highway 16. ❀ The store hasn't changed much over the years. It still has the original counters with glass-fronted bins that once held flour, meal, and beans. A modern scale was added in 1911 that Sue still uses. Taped to the side of the old scale are signed chits for items needed before payday. Customers still maintain charge accounts with the store, and their account cards are stored in an ancient wooden records box on the counter. ❀ After the store closes at 5:30 p.m., Sue delivers groceries to elderly residents who can't make it to the store. She's also the person the locals call when anything goes wrong. Snake in the house? She'll find someone to dispatch it! ❀ The floors have uneven, squeaky boards, the lights are dim, many of the canned goods are dusty, and the tables are well etched, but Cherokee Grocery and Sue Rhoades are living, working elements of Hill Country history. Stop in and enjoy the great food along with some fabulous stories.

101 N. Indian Street (State Hwy. 16) • (325) 622-4599

CHEROKEE

814, A Texas Bistro

Comfort

814, A Texas Bistro is a great little oasis of dining in the picturesque, sleepy little town of Comfort, where many structures are listed in the National Register of Historic Places and antique shopping is at its best. Owner Millard Kuykendall grew up on a Texas ranch and first experienced the restaurant business as a waiter in his younger years. As he worked his way up through the ranks, he discovered that he had a deep love of food and an affinity for cooking. After a job as a pastry cook, he concentrated on learning the basics and was eventually hired at Disney World in Orlando, Florida.

❈ When Millard and his family decided to return to the state and open their own restaurant, the historic post office building at 713 High Street was a perfect find. After a bit of remodeling and the addition of an outdoor dining area in the rear, 814 opened for business in 2006. The bistro serves a really good lunch, offering a varied selection of equally yummy dishes and a positively bodacious, finger-lickin'-good hamburger. It's one of the best burgers in the state. ❈ At dinnertime the menu switches into upscale mode with a selection of appetizers, salads, and entrées that changes weekly. Millard builds the menu around the freshest ingredients available, creating simple, honest tastes where fresh food is the star on the plate. "You're coming to my house when you come to 814 Bistro" is the way he sums up the dining experience here. Most of the diners are regulars, and on any given evening you'll find Millard popping out of the kitchen intermittently to chat with folks.

713 High Street •
(830) 995-4990 •
www.814atexas-bistro.com

JUMBO LUMP CRAB CAKES WITH CHIVE AIOLI

Yield: 8 crab cakes

1 large egg
1 teaspoon Worcestershire sauce
Juice from 1 large lemon
1 cup real mayonnaise
1 teaspoon Creole mustard
1 tablespoon minced flat-leaf parsley
1 pound jumbo lump crabmeat
Finely ground bread crumbs
Salt and black pepper to taste
Melted butter for sautéing

Chive Aioli

¼ cup chopped fresh chives
2 garlic cloves, peeled
½ cup water, or more as needed
1 cup real mayonnaise
Salt and black pepper to taste

Whisk the egg in a large mixing bowl until frothy. Add the Worcestershire sauce, lemon juice, mayonnaise, mustard, and parsley; whisk to blend well. Gently fold in the crabmeat, taking care not to break up the lumps. Add the bread crumbs a little at a time, just enough to bind the mixture together. Add salt and pepper to taste. Divide the crab mixture into 8 equal portions and pat into cakes.

Make the Chive Aioli. Combine the chives, garlic, and water in work bowl of food processor fitted with steel blade; process until smooth. Strain the puree through a fine strainer into a mixing bowl. Stir in the mayonnaise and whisk until smooth, adding additional water if the mixture is too thick. Season to taste with salt and pepper; set aside.

Sauté the crab cakes in melted butter over medium heat for 2 minutes per side, or until golden brown. Serve with the Chive Aioli.

(814, A Texas Bistro)

HIGH'S CHICKEN SALAD

The chicken salad at High's can become addicting, both as a salad and a sandwich filling on one of their yummy bread selections.

Yield: 4 sandwiches or salad servings

6 boneless, skinless chicken breasts, boiled, chilled, and cut into bite-size chunks
1 cup finely diced celery
½ cup thinly sliced green onions
1 cup red seedless grapes
¾ cup good-quality mayonnaise
¾ cup sour cream
Salt and freshly cracked black pepper to taste

Combine all ingredients in a large bowl and stir to blend well. Chill before serving.

(High's Café and Store)

HIGH'S HUMMUS

4 cups canned chick-peas, rinsed and drained
1 tablespoon minced garlic
½ teaspoon red (cayenne) pepper
½ teaspoon ground cumin
½ teaspoon paprika
1 teaspoon kosher salt
2 tablespoons minced flat-leaf parsley
¼ cup fresh lemon juice
2 tablespoons tahini paste
Olive oil
Toasted pita wedges
Greek olives, chopped tomatoes, olive oil, and minced parsley as garnishes

Combine all ingredients except olive oil in work bowl of food processor fitted with steel blade. Add just enough olive oil through the feed tube to get the mixture turning in the bowl (this is important for achieving a very smooth blend). Scrape down the sides of the bowl. As the mixture becomes smooth, slowly add more olive oil to achieve a silky smooth consistency. Process for about 2 to 3 minutes, almost "whipping" the hummus. Serve with toasted pita toasts, Greek olives, and chopped tomatoes. Drizzle olive oil on top and garnish with minced parsley.

(High's Café and Store)

High's Café and Store

Comfort

On Comfort's main shopping arcade, High Street, you'll find High's Café and Store, which offers fabulous coffee (also available in bulk) and coffee drinks. In the cozy interior or out on the front deck, you can enjoy irresistible breakfast pastries or lunchtime specialties such as made-from-scratch soups, salads with house dressings, sandwiches on artisan breads, and tantalizing desserts. Wines are served by the glass in addition to a selection of beers. ❀ High's also sells a great assortment of merchandise—kitchen items, toys and games, greeting cards, books on Texas subjects, pottery, Hill Country olive oil, and pickles and preserves. The impressive wine shop features wines from around the world and from local wineries. Wines produced by Cathie Winmill's Comfort Cellars Winery are sold only here and at the winery. ❀ Owners Brent Ault and Denise Rabalais started this special place with three sandwiches and some baked goods, along with their gourmet coffees. Brent's association with the food business goes back to his teen years when he and his sister opened an ambitiously large but successful deli in Fort Worth. He eventually moved to Fredericksburg and worked as a baker at some of the city's top restaurants. Denise, a native Houstonian, has a background in architecture and interior design and a passion for good food instilled by her Cajun-French father and Czech mother. After migrating to Fredericksburg, Denise managed the wine list for a large restaurant and met Brent. It's a partnership that works well. ❀ The pair settled on Comfort to start their business

because they loved its quirkiness and genuine small-town feel. Brent does the cooking in the tiny kitchen, and Denise tends the shop, always calling the regulars by name. High's was quickly embraced by the locals, who needed a place to get a good cup of coffee in the morning and a tasty lunch. Now the shopkeepers refer out-of-town customers, and the standing-room-only crowds tell it all.

726 High Street • (830) 995-4995 • www. highscafeandstore.com

Mandola Estate Winery and Trattoria Lisina

Driftwood

The Mandola Estate Winery, located south of Driftwood, brought a slice of Italy to the Texas Hill Country when it opened in early 2007. Damian Mandola and his wife, Trina, partnered with two longtime friends and fellow food and wine lovers, Stan and Lisa Duchman of Houston, to create the elegant and inviting winery. ❦ Damian is a well-known Texas restaurateur, with a career spanning over three decades. His ventures include Houston landmarks D'Amico's and Damian's Cucina Italiana as well as with the popular Carrabba's chain, which he founded with his nephew. ❦ The twenty acres of vineyards at Mandola Estates stretch over a slope above Onion Creek. With a climate similar to that of warm, dry southern Italy, Mandola focuses on growing grape varietals from Sardinia and Sicily. The plantings include such well-known varietals as pinot grigio, viognier, montepulciano, moscato, barbera, syrah, and sangiovese, along with lesser-known varietals like vermentino, nero d'Avilo, and aglianico. All seem to be thriving in the Hill Country. ❦ Winemaker and vineyard manager Mark Penna manages all the winery's operations and has designed a very impressive winery building with top-notch winemaking equipment. Before taking the helm at Mandola, Mark held various positions at other Texas wineries, including Cordier Estates in Fort Stockton, the state's largest winery. ❦ The winery features a spacious tasting room where visitors can sample the many Italian varietals produced. In addition to tasting the wines, you can browse a selection of Italian food products, cookbooks, and wine and kitchen accessories. ❦ Trattoria Lisina, the estate's restaurant, occupies a serene setting overlooking the winery, vineyard, and lush gardens. Reflecting the tempo of the Hill Country, the casual menu offers a varied selec-

tion of authentic country-style Italian foods, including Osso Buco, Veal Scaloppine, roasted duck, rabbit, house-made Italian sausages, pizzas, and pastas. In traditional Italian style, a meal at Trattoria Lisina usually consists of four courses.

13308 FM 150 West • (512) 858-1470 • www.mandola-wines.com • www.trattoriali-sina.com

CIPOLLINE IN AGRODOLCE

Yield: 6 to 8 servings

¼ cup extra-virgin olive oil
2½ pounds white cipolline onions, peeled
⅓ cup sugar
1¼ cups white wine vinegar
1¼ cups water
Salt and black pepper to taste

Combine all ingredients in a deep-sided 12-inch skillet. Cover and bring to a boil over medium-high heat. Reduce heat to a simmer and cook for 40 minutes, shaking the pan occasionally to keep the onions from sticking to the pan. When done, the onions should be easily pierced with a paring knife but not falling apart.

Remove the lid from the pan and continue to cook until the liquid has evaporated and the onions are glossy and dark brown, taking care not to burn them. Remove from heat and serve.

(Mandola Estate Winery and Trattoria Lisina)

SHRIMP DIABLO

Jumbo (16–20 count) shrimp, peeled, deveined, and butterflied

Onion slivers

Slices of seeded fresh jalapeños

Thin-sliced bacon

Salt Lick Dry Rub or your favorite rub

Salt Lick BBQ Sauce or your favorite barbecue sauce

Place an onion sliver and a slice of jalapeño inside each shrimp. Wrap the shrimp in bacon slices and season with dry rub. Place on hot grill and baste with barbecue sauce. Flip and baste again, cooking only until the shrimp turn white throughout. Serve hot, with additional sauce if desired.

(Scott Roberts)

The Salt Lick

Driftwood

Thurman Roberts Sr. and his wife, Hisako, founded this barbecue restaurant in 1969 on the ranch where he was born. Across the road from the ranch is Camp Ben McCullough, a reunion ground for Confederate veterans and their descendants. The Robertses figured that those folks might be potential customers for the family's renowned barbecue recipes. For their opening weekend they bought as much meat as they could afford and sold out. So the next weekend they bought a little more, and the rest, as they say, is history. ❀ In those early years the location of the Salt Lick was considered remote. But that has never stopped folks from making the journey—even though you have to bring your own beer and wine (Hays County is dry) and the restaurant takes cash only! ❀ The Salt Lick has expanded over the years, adding a separate pavilion for weddings and other events. But the original structure, built of stone quarried from the ranch, retains its rustic feel. Just inside the door you'll see one of the few open barbecue pits left in the Hill Country, piled high with brisket, ribs, and sausage. Meals are served family style on long picnic tables and include potato salad, coleslaw, beans, bread, pickles, and onions. On an average weekend the Salt Lick cooks 2,000 briskets alone! ❀ Now operated by the Robertses' son, Scott, the restaurant has won numerous "Best Barbecue" awards and has been featured in dozens of magazines, including *GQ, People,* and *Southern Living.* ❀ Barbecue, sauces and seasonings, and gift merchandise can be ordered through the Web site. Long-term plans for the Salt Lick include vineyards and, eventually, a winery. The dream continues.

18001 FM 1826 • (512) 858-4952 • www.saltlickbbq.com

Onion Creek Farm
Dripping Springs

Situated on thirty-six acres of mostly alluvial floodplain on the banks of Onion Creek in Dripping Springs, Onion Creek Farm has been certified by the Texas Department of Agriculture as an organic farm since 1983. ❈ Marianne Simmons, a botanist and former produce manager at Austin's original Whole Foods Market, purchased the farm in 1991 out of a deep love of produce and the business of produce. ❈ When Marianne took over the farm, she sold produce every Saturday morning at a farm stand she established in Dripping Springs. When the grocery giant H-E-B opened its flagship Central Market in Austin, Onion Creek Farm was the store's first direct vendor. She still sells to Central Market as well as to Austin's Wheatsville Co-op. ❈ Over the years, Marianne has transitioned from larger-scale, intensive vegetable production to growing and marketing a wide variety of culinary herbs, particularly basils, and certain seasonal vegetables, such as tomatoes, squashes, chilies, garlic, and greens. Nearly all the work on the farm is done the old-fashioned way with hoes and knives. All produce is hand-harvested at the peak of ripeness to deliver the fullest flavor and nutrition to the customer.

1611 Creek Road •
(512) 858-1090
• www.ocfarm.com

TOMATO AND BASIL SOUP

This soup, which combines three of Onion Creek Farm's products, tastes like the essence of summer itself, with the lusty notes of garlic, the zesty fresh flavor of juicy tomatoes, and the seductive nuances of fresh basil—three of the best things summer has to offer bundled up into one bowl!

Yield: About 3 quarts

9 large garlic cloves, peeled
2½ tablespoons minced basil leaves
2 small to medium onions, coarsely chopped
¼ cup extra-virgin olive oil
2½ quarts chopped garden-fresh Roma tomatoes and their juice
½ cup dry vermouth
2¼ teaspoons Oriental chili paste with garlic
1 tablespoon sugar
1½ quarts good-quality chicken stock
Salt to taste
3 cups whipping cream
½ cup grated Reggiano Parmesan cheese

Place steel blade in food processor. With processor running, drop the garlic cloves through the feed tube to mince. Stop the processor and scrape down the sides of the bowl. Add the basil and onions. Process until smooth and well pureed; set aside.

In a heavy 6-quart soup pot heat the olive oil over medium heat until hot. Add the garlic puree and cook, stirring often, until puree is wilted and transparent, about 5 minutes. Do not allow the mixture to brown. Puree the tomatoes with their liquid. Add the tomato puree to the soup pot and stir to blend well. Cook, stirring often, until the mixture is thickened, about 8 minutes. Stir in the vermouth, chili paste, sugar, and chicken stock. Blend well and add salt to taste. Cover pan and simmer for 30 minutes. Add the whipping cream and cook, covered, for 15 minutes.

Whisk in the Parmesan cheese, blending well. Simmer an additional 15 minutes. Serve hot.

(Terry Thompson-Anderson)

LAVENDER-INFUSED SANGRIA

Here are two great recipes from Sibby's collection at Onion Creek Kitchens, both paying homage to the bounty of lavender grown in the Hill Country.

Yield: About 1½ quarts

1 large orange, sliced
1 lime, sliced
1 lemon, sliced
Lavender Syrup (see recipe below)
¼ cup good-quality brandy
1 bottle (750 ml) red wine
1 bottle (12-ounce) club soda

Lavender Syrup

1 cup water
1 cup sugar
2 tablespoons dried culinary lavender buds

Make the Lavender Syrup. In a small saucepan, bring water and sugar to a boil and stir until dissolved. Remove from heat and add lavender buds. Cover and allow syrup to stand for 1 hour. Strain into a glass container, cover, and keep refrigerated for up to a few weeks.

In a large glass pitcher, combine the fruit slices. Add the brandy, Lavender Syrup, and wine. Stir to blend well. Refrigerate until well chilled, about 2 hours. When ready to serve, fill wine glasses with ice and fill glasses one-half to two-thirds full of the sangria mixture. Top with desired amount of club soda. Add fruit slices to glasses.

(Onion Creek Kitchens)

Onion Creek Kitchens at Juniper Hills Farm

Dripping Springs

Travis "Sibby" Barrett is a native Texan who has spent most of her life in Dallas, successfully operating a flower company, a bakery and cake shop, an event-planning business, and a restaurant. To "slow down a bit," Sibby packed up and headed for the Hill Country, where she established Juniper Hills Farm southwest of Dripping Springs. ❖ The farm features four casitas and two studios available for a quiet respite in a scenic, restful spot. A gorgeous infinity-edge pool completes the scene for an over-the-top weekend experience.

In her lovely stone home Sibby offers her wildly popular Onion Creek Kitchens Cooking School, where reservations for her "hands-on" classes are often booked for months in advance. ❖ From the frenzy of farmer's market, cheese maker, herb and olive orchard shopping, mincing, slicing, dicing, whisking, and folding, to the chaos of sautéing, poaching, grilling, and roasting, students work up a healthy appetite while Sibby adds the magic touch that brings it all together in a beautiful, utterly delicious feast for both the eyes and the palate. ❖ Delighted students fill their plates with the mouth-watering fruits of their labor and head to the banquet-size table in the dining area to enjoy the meal and discuss the many new tips and techniques they have learned.

5818 RR 165 • (830) 833-0910 • www.juniperhillsfarm.com

LAVENDER-ORANGE ICE CREAM

2½ cups whole milk

2½ cups whipping cream

2 vanilla beans

Zest of 1 orange, 1 teaspoon reserved

6 or 7 fresh lavender sprigs, 1 sprig reserved, or substitute 1 tablespoon plus 1 teaspoon dried culinary lavender

6 egg yolks

1 cup sugar

Juice of 1 orange

In a large saucepan combine milk and cream. Split vanilla beans lengthwise. Using a small paring knife, scrape out seeds and add the seeds and pods to the milk mixture. Add the orange zest and lavender sprigs. Bring to a boil over medium heat. Remove from heat and allow to steep for 30 minutes.

Meanwhile, finely chop the remaining lavender sprig and set aside with the reserved 1 teaspoon of orange zest. Combine egg yolks and sugar in bowl of electric mixer. Beat at medium speed until mixture is light lemon-yellow in color and the sugar is completely dissolved. Reheat the milk mixture until very hot but not boiling. Slowly pour about a third of it into the egg yolks, beating constantly. Then pour the milk and egg mixture into the saucepan with the remaining milk. Cook over medium heat, stirring constantly with a wooden spoon, until the custard is thick and coats the back of the spoon. Do not let it boil. Remove the vanilla bean pods and strain the mixture into a large bowl. Add the reserved lavender and orange zest and whisk in the orange juice. Allow mixture to cool to room temperature, stirring occasionally.

Refrigerate until well chilled, or up to two days. When ready to make the ice cream, process in an ice cream freezer according to the manufacturer's instructions. Serve when firm enough to scoop.

(Onion Creek Kitchens)

CILANTRO AND GOAT CHEESE MASHED POTATOES

For a different twist on plain old mashed potatoes, try this variation. Even those who claim to not care for the taste of goat cheese love 'em!

Yield: 4 servings

1 pound small unpeeled new potatoes, quartered
2 tablespoons unsalted butter
1 small onion finely diced (about ½ cup)
4 ounces Pure Luck Texas Plain Chèvre
1 heaping tablespoon minced cilantro
¼ teaspoon freshly ground black pepper
¾ teaspoon sea salt

Place potatoes in a heavy 3-quart saucepan and add cold water to cover. Bring to a full, rolling boil over medium-high heat and cook for 20 minutes, or until potatoes are very soft.

While potatoes are cooking, melt butter in a heavy 8-inch skillet over medium heat. Add onions and sauté, stirring often, until wilted and transparent, about 5 minutes. Remove from heat and set aside.

Drain potatoes in a colander and transfer to a medium bowl. Add the sautéed onions and any residual butter in the skillet, chèvre, cilantro, salt, and pepper. Mash the potatoes with a potato masher, incorporating the other ingredients. Leave them slightly lumpy. Serve hot.

(Terry Thompson-Anderson)

Pure Luck Texas
Dripping Springs

The Pure Luck Farm and Dairy, home of 100 Nubian and Alpine goats and five acres of organic farmland on Barton Creek, produces artisan goat cheese and certified organic culinary herbs. Every aspect of the operation is done by hand with love. ❉ Pure Luck was founded by the late Sara Sweetser Bolton in 1979 on acreage that had been a working tomato farm in the 1930s. She tilled the rich soil and put in a productive vegetable garden to feed her two young daughters, Gitana and Amelia. While caring for a friend's goats, she decided to get a herd of her own and was soon producing goat's milk and cheese. In 1988 Pure Luck was one of the first farms in Texas to apply to the Texas Department of Agriculture for certification as an organic farm. ❉ The first crop grown for sale on the farm was cucumbers, which were eagerly purchased by Whole Foods Market in Austin. The farm was soon growing fresh-cut, organic culinary herbs, which are now harvested twice a week and delivered to local Whole Foods Market and Central Market stores. The cheese production went commercial in 1995. Continuing the labor of love created by their mother, Amelia serves as

president of Pure Luck and Gitana as vice-president. The award-winning cheeses are made with the assistance of cheese maker Juana Mora. ❉ As a farmstead operation, Pure Luck uses only milk produced on the premises in making its six distinct types of goat cheese: various seasoned chèvres, feta, Del Cielo (a Camembert-type cheese), Sainte Maure, Hopelessly Bleu, and Claire de Lune (a hard aged cheese). ❉ The cheeses are served in upscale Central Texas restaurants and can also be purchased in Austin at Boggy Creek Farm, Central Market, Farm to Market Grocery, Wheatsville Co-op, and Whole Foods Market as well as at Becker Vineyards near Stonewall and Texas Specialty Cut Flowers in Wimberley. Private tours, which include tastings of the various cheeses, can be arranged at the farm.

101 Twin Oaks Trail • (512) 858-7034
• www.purelucktexas.com

Rolling in Thyme and Dough

Dripping Springs

One of the Hill Country's most unique places is Rolling in Thyme and Dough, a bakery and café ensconced in a 100-year-old cottage on a large tree-shaded lot along the south side of U.S. 290. Take a few steps inside the door and you are aware that this is a local hangout, filled with ladies chatting over breakfast pastries and coffee. Nooks and crannies are filled with aromatic baked goods. ❧ Owners Marsha Shortwebb and Fabienne Bollom have created a special space, offering home-baked pastries, tarts, pies, homemade whole-grain breads, and great lunches—all made on the premises from fresh ingredients and organic produce. Marsha does the cake decorating, creating some very impressive cakes, including a few with Texas wildflower designs. ❧ The menu, which includes soups, salads, pizzas, sandwiches, wraps, and daily specials, reflects the cooking style of Belgian-born Fabienne, who likes to blend European recipes with American cooking methods. ❧ You can enjoy a relaxing cup of coffee, specialty coffee drinks, and hot teas, either outside in the garden or inside. Marsha and Fabienne are deeply involved in community support, featuring the works of local artists in their shop and hosting such events as Art in the Garden parties. Peruse the selection of handmade items (such as purses, soaps, cards, and candies) as well as the seasonal herb plants available in the gardens. ❧ Opened in 2005, the cozy eatery has been enthusiastically embraced by the Dripping Springs community. Few could resist the heavenly smells emanating from the tiny kitchen or the ambiance created by the collection of odd tables and chairs alongside antique sideboards, armoires, pie safes, and hutches filled with tempting goodies.

333 U.S. 290 West •
(512) 894-0001

TARTE À LA FRANGIPANE
(Almond Tart)

The almond meal used in the filling for this recipe is made by grinding 8 ounces (2 cups) of sliced, skin-on almonds in the food processor until they are reduced to a fine meal.

Yield: 6 to 8 servings

1 package frozen Pepperidge Farm puff pastry (2 sheets), thawed

Almond Filling
½ pound (2 sticks) butter
1 cup sugar
1 cup almond meal
1 cup all-purpose flour
3 eggs plus 1 egg for egg wash

Glaze
4 ounces apricot preserves
½ cup powdered sugar
1 teaspoon fresh lemon juice
Toasted sliced almonds

Preheat oven to 350 degrees. Set aside one sheet of the puff pastry. Roll out the remaining sheet until very thin. Place the pastry in a 12-inch tart pan, trimming off the edges. Refrigerate.

Make the Almond Filling. In bowl of electric mixer, beat the butter at medium speed until very creamy. Add the sugar and 1 egg, beating to blend well. Add the almond meal and the remaining 2 eggs, one at a time, stopping to scrape down the sides of the bowl after each addition. Beat just to blend, then add the flour, beating until well blended.

Turn the filling out into the chilled pastry, spreading evenly. Using a sharp knife, cut the remaining sheet of puff pastry into strips ½ inch wide. Arrange the strips in a lattice pattern over the top of the filling. Beat the eggs for the egg glaze and brush it over the surface of the lattice strips. Bake tart in preheated oven for 35 to 40 minutes, or until filling is set and pastry is golden brown.

While the tart is baking, heat the apricot preserves until very thin. Strain through a fine strainer, stirring to remove all of the jelly; discard the pulpy fruit. When tart is done, brush the preserves over the top. Combine the powdered sugar and lemon juice, blending well. Brush over the top of the apricot glaze, then scatter some sliced almonds over the tart. Allow glaze to set before slicing.

(Rolling in Thyme and Dough)

Alamo Springs General Store and Café

Fredericksburg

The Alamo Springs General Store and Café is definitely off the beaten path—about ten miles southeast of Fredericksburg via the Old San Antonio Road. Look for signs to the Old Tunnel Wildlife Management Area, where three million bats make their home in an abandoned railroad tunnel. ❧ The "Bat Cave," as the tunnel is called, is the main reason why Mike Tangman opened the café. He figured the folks coming to see the bats would be made-to-order customers, as there certainly isn't anywhere else nearby to get a bite to eat! He was right, but his clientele is much larger. Residents from all over the area hang out here, and it's a favorite stop for bicyclists riding the back roads. It's been written up in publications all over the country and memorialized by several film crews. You can even buy t-shirts and tank tops with the café's mantra: "Alamo Springs Café—Where Anybody Is Nobody." ❧

Mike opened the store and café on Memorial Day in 2004. His original concept was to sell upscale deli-style sandwiches and homemade soup in the café, with a burger to satisfy kids. The very first customer to walk in the door ordered the burger and raved about how good it was. So the next customer ordered one too, and the next . . . . At the end of his first week, Mike threw away 80 pounds of gourmet deli meats and ordered 80 pounds of ground chuck.❧ This is a seriously delicious burger, with awesome hand-cut fries, and it's been named the "Best Burger in Texas" in news articles from coast to coast. You have a choice of great buns—Sourdough Jalapeño-Cheese or Whole Wheat. Mike's burger reigns, although great sandwiches (like Chicken Chipotle), salads, and homemade potato chips are on the menu. ❧ To spend more time managing the store, Mike hired Chef George Fuentes to take over the kitchen and add dinner specials, such as Grilled Quail

Salad. ❧ This is a fun place. You can drink a brew or two—you'll find your Lone Star as well as plenty of Shiner Bock and Negro Modelo—or enjoy wines by the glass or bottle. There's a front porch and a big deck, offering perfect venues for people-watching.

107 Alamo Road
• **(830) 990-8004**

BLACKENED CENTER-CUT PORK LOIN CHOP

Chef George Fuentes serves this delectable pork chop with a citrus and mango rice studded with toasted almonds and basil.

Yield: 4 servings

4 (8-ounce) center-cut boneless pork loin chops
Your favorite Cajun-style seasoning

Sauce

¾ cup dry white wine
¼ cup honey
¼ cup plus 1 tablespoon Dijon-style mustard
⅔ cup whipping cream
¼ teaspoon salt
¼ teaspoon ground white pepper
1 tablespoon butter

To make the sauce, combine the wine, honey, and mustard in a heavy saucepan and bring to a full boil. Lower heat and cook until the mixture is reduced by half. Stir in the whipping cream and season to taste with salt and white pepper. Simmer for an additional 15 minutes.

While the sauce is simmering, liberally season both side of the pork chops with Cajun seasoning, patting the seasoning into the meat with your fingers. Heat a cast-iron skillet over medium-high heat until very hot. Add the pork chops and cook until an instant-read thermometer inserted into the thickest part of the meat registers 145 degrees. Set aside and keep warm.

Remove sauce from heat and whisk in the butter. Serve the chops topped with a portion of the sauce.

(Alamo Springs General Store and Café)

Yield: 4 servings

½ pound backfin crabmeat
⅓ cup chopped parsley
½ teaspoon salt
Pinch of black pepper
Juice of 1 lime
Olive oil
1½-inch PVC pipe, cut into four 4-inch lengths
(for molds)
3 medium tomatoes, chopped
2 small avocados, chopped
Crostini
Classic Remoulade

Crostini

Half of a baguette, sliced into ¼-inch pieces
Olive oil
Pinch of salt
Parsley

Classic Remoulade

1 cup of mayonnaise
¼ cup chili sauce
2 tablespoons Creole mustard
2 tablespoons extra-virgin olive oil
1 tablespoon Louisiana-style hot sauce, or to taste
2 tablespoons fresh lemon juice
1 teaspoon Worcestershire sauce
4 medium scallions, chopped
2 tablespoons chopped fresh parsley
2 tablespoons chopped dill pickle
1 clove garlic, minced
½ teaspoon chili powder
1 teaspoon salt, or to taste
½ teaspoon ground black pepper
1 teaspoon capers, chopped (optional)

To make the Classic Remoulade, combine all ingredients in work bowl of food processor fitted with steel blade. Process until smooth. Refrigerate until ready to use.

Preheat oven to 350 degrees. Toss baguette slices with olive oil, salt, and parsley. Bake in preheated oven 7 to 10 minutes, or until crisp but not browned.

Combine crabmeat and half the parsley in a medium bowl. Add salt, pepper, and lime juice; mix well. Oil the inside

(continued)

August E's

Fredericksburg

When word spread that a new restaurant by the name of August E's was opening in a rustic log cabin just east of Fredericksburg, the town was abuzz with speculation. "What kind of name is that?" everyone asked. ❄ By the time Dawn and Leu Savanh opened August E's in October of 2004, it was common knowledge that they had also purchased the historic August Eber homestead in nearby Grapetown and named their restaurant in honor of the Eber family's role in area development. Mystery solved. ❄ In 2007 Dawn and Leu were able to purchase a large building in the heart of Fredericksburg's historic district. They converted the interior into a space that wows—the perfect setting for their menu, which they've labeled Nouveau Texas Cuisine. It's an innovative fusion of traditional Texas fare with the foods of Leu's Laos/Thai roots. The total package of August E's, from the décor to the food to the superlative level of service, has added a new dimension of sophistication to dining in Fredericksburg. August E's garnered a Three Diamond rating from AAA, and in 2008 it became only the seventeenth restaurant in the state to make the prestigious DiRoNA list (Distinguished Restaurants of North America). ❄ Dawn began her career in hotel destination management, eventually working for the Anatole Hotel in Dallas. Here she met Leu, a martial arts instructor who had immigrated to the United States when he was fourteen and spoke not a word of English. After discovering that Leu was a fabulous cook, Dawn introduced him to the Anatole's chef, who also saw his potential and brought him into the hotel's kitchens to train as a chef. When Dawn and Leu were ready to try a restaurant of their own, they relocated to an area they had visited often and grown to love. ❄ Leu says the inspiration for his fabulous flavor pairings comes from a creative spot deep inside, adding, "When you first love to cook, the rest comes naturally." His creativity leads to lots of experimenting, always starting with the freshest and best local ingredients he can find. The Savanhs hope to see the day when the Hill Country has a large co-op of local farmers and ranchers growing organic vegetables and free-range animals for area restaurants.

203 E. San Antonio Street • (830) 997-1585 • www.august-es.com

of each length of pipe and stand it upright on a cutting board or plate. Fill a third of the pipe with the crab mixture, then add chopped tomatoes and avocados until the pipe is half full. Repeat with the remaining crab mixture, tomatoes, and avocados; top with the remaining chopped parsley and set aside.

Place the Classic Remoulade in a plastic squeeze bottle. (You may have to cut off a portion of the pointed tip to allow the sauce to pass through.) Squiggle a liberal portion of the sauce across each individual serving plate. Using a small, clean glass bottle, press the crab mixture out of the pipe and onto the center of a serving plate. Arrange 4 to 6 Crostini around each Crab Stack and serve.

Variation: Instead of the remoulade, serve with wasabi aioli, which can be purchased commercially at Whole Foods Market or an Asian market.

(Leu Savanh)

GRILLED JALAPEÑO-STUFFED QUAIL WITH HILL COUNTRY STRAWBERRY-CABERNET GLACÉ AND MESCLUN GREENS WITH WALNUT DRESSING

Mesclun is a salad mix of small, young salad leaves and is found in specialty produce markets and many upscale supermarkets. Although the assortment of greens can vary, the salad mix is generally composed of arugula, dandelion, frisée, mizuma, oak leaf, mâche, radicchio, and sorrel. When selecting mesclun, look for crisp leaves with no sign of wilting.

Yield: 2 servings

½ cup Becker Vineyards cabernet sauvignon
½ cup sugar
1 shallot, peeled and chopped
2 whole cloves
1 teaspoon whole black peppercorns
¼ cup strawberry preserves
4 semi-boneless quail
2 fresh jalapeños, split lengthwise
4 strips of applewood-smoked bacon
Kosher salt and freshly ground black pepper

Mesclun Greens with Walnut Dressing

2 tablespoons walnut oil
¼ cup olive oil
1 tablespoon freshly squeezed lemon juice
2 tablespoons toasted walnut pieces, chopped fine
Kosher salt and freshly ground black pepper
2 cups mesclun greens

Combine the wine, sugar, shallot, cloves, and peppercorns in a medium saucepan over medium-high heat; bring to a boil. Reduce in volume until the wine mixture begins to lightly coat the back of a spoon. Remove from heat and strain while still hot; discard the spices. Add the strawberry preserves to the hot wine mixture and stir until preserves are melted and well mixed; set aside and keep warm.

Preheat gas grill. Place a jalapeño half in the cavity of each quail and wrap quail with a strip of bacon, securing with a toothpick. Season quail with salt and pepper and grill for about 5 minutes per side, turning once, or until the juices from the quail run clear. Transfer quail to a warm platter and remove toothpicks.

Whisk together the walnut oil, olive oil, lemon juice, and walnut pieces; add salt and pepper. Toss the mesclun greens with the walnut dressing.

Place the greens in the center of the platter. Arrange quail around the platter, slightly overlapping the greens. Pour the warm glacé over the quail and serve at once.

The Cabernet Grill—Texas Wine Country Restaurant

Fredericksburg

The Cabernet Grill, originally called the Cotton Gin, has been constantly evolving since its opening in 2001 as part of a lushly landscaped bed and breakfast compound, complete with authentic log cabins. Dating from the mid-1800s, the cabins were found in an area of Kentucky and Tennessee, disassembled, and moved to their present location, where they were painstakingly reassembled with modern amenities and antique furnishings added. The restaurant failed after only six months in operation, and the entire property was placed on the market. ❋ Ross Burtwell fell in love with the place and purchased the entire compound. He received his culinary training at El Centro College in Dallas and had the opportunity to hone his skills under some top-rated chefs in the Westin Hotel chain. He was working in San Antonio when he met his wife, Marianna, whom he refers to as the "dynamo who doesn't like the spotlight." Marianna does all of the lunch preparation at the Cabernet Grill, and it's one of Fredericksburg's most popular lunch spots. It's a good idea to make reservations for dinner to avoid a potentially long wait. ❋ Ross has slowly developed his menu, offering a great cross section of foods reflecting the way we eat in Texas today, with roots in

the cuisine of the early Texas settlers. Each dish shows his passion for food as well as his attention to detail, weaving flavors that pair perfectly with Texas wines. In fact, he changed his wine list to an all-Texas wine list, making Cabernet Grill truly a Texas Wine Country Restaurant! Ross and his longtime manager, Mike Boase, constantly seek out new Texas wines. Before a winery's products are added to the list, Ross takes his entire waitstaff to the winery to become fully acquainted with its wines, its history, and its owners. As a result, the staff is very excited about the wine program and able to assist diners with great finesse in selecting the perfect wine to pair with the dishes they have ordered. The Cabernet Grill is a totally Texas experience right in the heart of the Hill Country.

Cotton Gin Village • 2805 S. State Highway 16 • (830) 990-5734 • www.cottonginlodging.com

Chocolat

Fredericksburg

Master chocolatier Lecia Duke first learned the art of making candy from her grandmother and then attended the famous Wilton School of Cake Decorating and Confectionary Art in Chicago. She started her chocolate career by making corporate logos in chocolate. Some of her early customers were actress Mary Tyler Moore and the Judds of country music fame. ❄ After apprenticing under a Swiss chocolatier to learn how to make chocolates with liquid centers, Lecia founded Quintessential Chocolates in 1984. Her company became the sole U.S. producer of chocolates made with the fine confectionary process known the Liquid Moisture Barrier Technique. ❄ In 2002 Lecia opened her present shop, Chocolat, in a lovely old building on Fredericksburg's West Main Street. ❄ To add an American twist to the process, Lecia began to fill her chocolates with a wide spectrum of wine-based and nonalcoholic liquid centers in addition to the traditional liqueurs. And even her liqueurs aren't traditional. ❄ The selection of Premium Spirit Chocolates includes morsels filled with chocolate liqueur, crème de menthe, whiskey, Kentucky bourbon, peach schnapps, Russian vodka, Ohranj (orange), spiced rum, Tennessee whiskey, toffee liqueur, and more. The Crema d'Almendrado Tequila chocolates are notable for having their three distinctive flavors: almond, tequila, and chocolate. In ten years the company has made over 115 different flavors of liquid-center fillings. ❄ In addition to chocolates filled with cabernet, Cognac, tawny port, and raspberry wine, the selection features chocolates made with wines from several private-label producers, including two Texas wineries. ❄ Chocolat also makes handmade European-style truffles, barks, nut clusters, and other fine specialty confections. All candies are made fresh daily on the premises. ❄ If your Hill Country itinerary includes Fredericksburg, be sure to visit Chocolat. You can even watch the chocolates being made!

330 W. Main Street • (830) 990-9382 • www.chocolat-tx.us

Cranky Frank's Barbecue Company

Fredericksburg

Dan and Kala Martin opened Cranky Frank's in 2003. Their son Frank, only a year old at the time, was always cranky, hence the name of their barbecue joint. ❄ But Cranky Frank's is more than a joint. There's a nice dining room in addition to picnic-style seating outdoors, where you can smell the meat slow-cooking over mesquite coals in huge pits. ❄ The most important thing, though, is the excellent barbecue: perfectly smoked brisket with a nice little red smoke ring and pork ribs that fall off the bone with a couple of tugs from the teeth. There is also great sausage, pork roast, and chopped beef with all the requisite sides—potato salad, pinto beans, coleslaw, burger pickles, and chopped onions. Everything's made fresh daily, and it's all good. The Martins make two kinds of barbecue sauce, one from an old German family recipe and another in a traditional style. There are plenty of cold longnecks to savor with your meal. ❄ When you're hankering for some good barbecue while you're in Fredericksburg, stop by Cranky Frank's and step up to the counter to place your order.

1679 S. Washington Street (U.S. 87 South) • (830) 997-2353

Der Küchen Laden

Fredericksburg

In search of a microplane or perhaps a kugel-hopf pan, a Japanese hand-rolled and folded cobalt steel boning knife, some great coffee beans, an obscure cookbook, or some French porcelain ramekins? You can find them all at Der Küchen Laden (The Kitchen Shop). The store is a mecca for confirmed foodies—and an interesting browse even for those who aren't serious cooks. The great selection includes kitchen towels, potholders, and decorative kitchen accessories as well as the specialized, hard-to-find items.

❄ Proprietors Jerry and Penny Perry-Hughes, Fredericksburg natives, both grew up cooking. After graduating from a prep school whose Austrian headmaster taught cooking classes, Penny attended Le Cordon Bleu in London for four months; she went back for more training in 1976. Jerry is noted for his in-store demonstrations and his knowledge of fine cookware. ❄ Der Küchen Laden first opened in Fredericksburg in 1978 as a very small kitchenwares shop. Jerry and Penny bought the shop in 1993 and in 2000 moved it to its present location, the ground floor of the historic Keidel Memorial Hospital. Penny's grandfather, a physician, had bought the building in the 1930s and turned it into Fredericksburg's first modern hospital after it had been operated as a retail store. Now the larger rooms as well as the various nooks and crannies are filled with anything and everything needed in a kitchen. ❄ Jerry and

Penny see the shop as offering possibilities to cooks, empowering them to cook good food at home. One of the best things about Der Küchen Laden is that the staff is very knowledgeable about the merchandise. If you can describe what you're looking for, they can most likely lead you to it—and show you how to use it!

258 E. Main Street • (830) 997-4937 • www.littlechef.com

Der Lindenbaum

Fredericksburg

Amid the many tourist-oriented German restaurants in Fredericksburg, one in particular is a shining example of a place serving authentic German food. Ingrid Hohmann, a professional chef, opened Der Lindenbaum with a commitment to serving the real deal. She named the restaurant after Germany's beloved linden tree, which grew in front of her family's home in the Eifel Mountains. ❄ A graduate of the prestigious hotel and restaurant school at Maria Laach, Ingrid began her professional career as a *konditormeister*, or pastry chef, making apple strudels, Black Forest cake, and other European pastries. ❄ Der Lindenbaum offers many dishes not commonly found on German restaurant menus in the United States. One of the specialties here is Königsberger Klopse, a fabulous dish of beef and pork meatballs swimming in a delicate caper sauce over egg noodles. Other excellent dishes are Rheinischer Sauerbraten, beef roast marinated for an entire week in a sweet-sour sauce and then slow-roasted until it's fork-tender; Curry Huhn, an amazing curried chicken dish; Gulasch, a spicy beef stew; Schweinekotelett, pork chops in a heavenly mustard sauce . . . the list goes on! Der Lindenbaum's schnitzels are fantastic, as are the German sausages. ❄ You'll find one of the Hill Country's best selections of German beer and wine at Der Lindenbaum—requisite beverages for Ingrid's food! Naturally, you'll want to save room for those sinfully rich German desserts with which she began her culinary career and has honed to perfection over the years. You'll also enjoy the cozy atmosphere of the intimate rock-walled dining room.

312 E. Main Street • (830) 997-9126 •
www.derlindenbaum.com

SERBISCHE BOHNEN SUPPE
(Serbian Bean Soup)

Ingrid serves this comfort-food soup in shallow soup plates. It's easy to prepare, but don't rush it. The secret is in the slow simmering. Serve it with a nice dry German Riesling and some dark bread with real European-style butter. Home-cooked beans and homemade beef stock are preferable, but canned will suffice.

Yield: 4 servings

3 cups Great Northern beans and their cooking liquid
1 large yellow onion, finely chopped
2 green bell peppers, finely chopped
2 red bell peppers, finely chopped
2 cups finely chopped Polish sausage
8 cups beef stock
Tabasco to taste
Salt and freshly ground white pepper to taste
1 tablespoon good-quality soy sauce
2 pork rib bones
2 beef rib bones

Combine all ingredients in a heavy, thick-bottomed 8-quart soup pot over medium heat, stirring to blend well. Simmer the soup slowly for 2 hours, stirring occasionally, or until the beans are quite soft and a thick gravy has formed. Remove the rib bones (great for behind-the-scenes munching in the kitchen). Ladle the soup into shallow bowls and serve hot.

(Der Lindenbaum)

Fredericksburg Brewing Company

Fredericksburg

The Fredericksburg Brewing Company, owned by Richard and Rosemary Estenson, opened in 1994 after the state passed legislation allowing the operation of brewpubs. It is the oldest brewpub in Texas and has been a favorite gathering place since the beginning. The brewery is located in a renovated 1890s limestone building in the heart of the town's busy shopping district. ❄ Before opening the brewery, the Estensons researched brewing equipment and processes throughout the United States as well as in Germany, Hungary, and the Bohemia region of the Czech Republic. The beers are brewed according to true German traditions, with a current storage capacity of 7,500 gallons. Plans are under way for adding an additional 2,000 gallons of storage in 2008. The brewing equipment—huge, shining copper and stainless steel tanks—is in full view of the long bar and dining area. ❄ Brewer Rick Green and his assistant, Alton Huebner, create five styles of beer in seasonal brews. You can even buy the beers in half-gallon jugs to take home! ❄ At the 2007 Great American Beer Festival in Denver, the largest beer competition in the nation, the brewery's Pioneer Porter received a gold medal in the Brown Porter category. Enchanted Rock Red Ale, a dark English-style ale, won the bronze medal in the Irish-Style Red Ale category. These are two of the brewery's most popular beers.

245 E. Main Street • (830) 997-1646 • www.your-brewery.com

Fredericksburg Grass-Fed Beef

Fredericksburg

Fredericksburg Grass-Fed Beef is owned by Chuck and Teppi Schmidt and Lonnie and Patricia Marquardt. Chuck and Patricia are siblings, the third generation to operate the family ranch, SMS Ranch. The idea for organic, pesticide-free pastures on the ranch began with their dad, who viewed ranching from a holistic approach. ❉ The herd, consisting primarily of Hereford-Angus crosses, is wholly family owned, and many are descended from the first cows brought to the ranch by Chuck and Patricia's grandparents over eighty years ago. The Schmidts and Marquardts have learned that docile cows tend to produce tender meat, so a heifer's disposition is an important criterion when adding to the herd. ❉ The native pastures where the calves are born, raised, and finished have not been treated with pesticides or herbicides in over fifteen years. Native alfalfa grass is used as protein feed, and no supplemental grain is fed. The health of the herd is maintained by frequent pasture rotation. This means that the animals have a constant supply of a broad variety of plants from which to feed, producing a healthier animal with more flavorful meat. No antibiotics or hormones are used. ❉ Processing is done in a USDA-inspected packing house in Fredericksburg. All meat is frozen before delivery to ensure freshness. ❉ The Schmidts and Marquardts are dedicated to organic, sustainable ranching principles, although the ranch is not yet certified organic. They are striving hard to educate consumers about the benefits of eating grass-fed beef. Chuck states, "Eating grass-fed beef just makes sense in light of today's health concerns. The beef is lower in fat and contains two to six times more omega-3 fatty acid than grain-fed beef and also three to five times

more conjugated linoleic acid." ❉ Fredericksburg Grass-Fed Beef products are marketed through the Web site, and the company will deliver within a seventy-mile radius of Fredericksburg.

(830) 990-9353 • www.fredericksburg-grass-fed-beef.com

ROSEMARY-ORANGE RUM CAKE

This cake is one that has been around since the herb farm first opened and is still popular today.

Yield: One Bundt cake

1 package two-layer yellow cake mix

1 small package vanilla instant pudding mix

1 tablespoon finely minced rosemary

Grated zest of 1 orange

½ cup water

½ cup canola oil

½ cup light rum

4 extra-large eggs

1 cup chopped, toasted pecans

Rosemary sprigs for garnish, if desired

Glorious Glaze

¼ pound (1 stick) unsalted butter

1 cup sugar

¼ cup water

¼ cup rum

Preheat oven to 325 degrees. Spray a 10-inch Bundt pan with nonstick baking spray. Set aside. Combine the cake mix, pudding mix, rosemary, and orange zest in work bowl of food processor fitted with steel blade; process until well mixed. Add the water, canola oil, and rum; process to blend well. Add the eggs, one at a time, stopping to scrape down the sides of the bowl and processing just to blend after each addition. Turn mixture out into a bowl and stir in the toasted pecans. Pour the batter into the prepared Bundt pan and bake for 1 hour, or until a wooden toothpick inserted in center of cake comes out clean.

While the cake is baking, prepare the Glorious Glaze. Combine all ingredients in a heavy 2-quart saucepan and bring to a boil. Boil until the mixture reaches the soft-ball stage (235 degrees on a candy thermometer).

Pour the glaze over the cake as soon as it comes out of the oven. Place cake on a wire rack to cool, allowing the glaze to soak in completely before removing cake from pan. Invert the cake onto a platter to serve. Garnish servings with rosemary sprigs, if desired.

(Fredericksburg Herb Farm)

Fredericksburg Herb Farm

Fredericksburg

Like many Hill Country ventures, Fredericksburg Herb Farm started when Bill and Sylvia Varney had an opportunity to leave their corporate jobs in Houston and escape to a laid-back life in Fredericksburg, where Bill was offered a job as manager of a nursery. ❀ In 1985 the Varneys opened a small shop on Main Street where they sold herbal products. Out back they grew herbs and flowers, from which they produced vinegars, lotions, and gels. They also added a tearoom where they served herbal teas and herb-laced desserts. Business exploded when the shop was featured in *Victoria* magazine and when the Varneys' Edible Flowers Vinegar won first place at the 1991 International Fancy Food Show in New York. ❀ In 1991 they purchased an abandoned four-acre farm with a small limestone homestead and named the new venture Fredericksburg Herb Farm. Eventually the farmhouse became the Herb Farm Restaurant, serving delicious fare using the bounty of the farm's edible herbs and flowers. The Varneys were invited twice to prepare dinners for the James Beard Foundation. ❀ Today the herb farm consists of six lovely theme gardens, a day spa, a bed and breakfast, the restaurant, and a retail shop. Visitors can stroll through the gardens, which offer relaxing places to stop and enjoy the serenity and surrounding aromas. The patio of the restaurant has tables where you can enjoy a cup of herbal tea or a glass of wine and perhaps a dessert or light appetizer. Many varieties of herb plants are available for purchase.

407 Whitney • (830) 997-8615 • www.fredericksburgherbfarm. com

VARNEY'S-CHEMIST-LADEN

A FINE
FRAGRANCE, TOILETRIES
& HERB SHOP
FOR
LADIES AND GENTLEMEN

Toiletries

OILS

FREDERICKSBU...
HERB FARM

Natural Toiletr...
Herbal Body Ca...
Gourmet Condi...
Essential Oils
Books
Garden Ornam...

BAKED CHÈVRE CHEESECAKE

This delicious savory cheesecake featuring goat cheese is a Hilltop Café classic.

Yield: 8 to 10 servings

Crust

2½ cups fresh bread crumbs

3 tablespoons clarified butter

3 tablespoons grated Romano cheese

2 garlic cloves, minced

2 teaspoons chili powder

1 medium pear, peeled, cored, and finely diced

Filling

28 ounces cream cheese at room temperature

8 ounces soft chèvre (1 cup)

⅓ cup half-and-half

1 cup sour cream

2 garlic cloves, roasted

¼ cup finely diced sun-dried tomatoes

2 tablespoons minced basil

2 tablespoons minced oregano

2 tablespoons minced marjoram

½ teaspoon kosher salt

4 eggs

Sour cream, rosemary sprigs, and champagne grapes for garnish (optional)

To make the crust, combine all ingredients in work bowl of food processor and process until smooth and well blended. Press the crust into the bottom and sides of a 9-inch spring-form pan. Refrigerate for at least 30 minutes.

Preheat oven to 350 degrees. To make the filling, combine the cream cheese, chèvre, half-and-half, and sour cream in work bowl of food processor. Process until smooth. Add the garlic, sun-dried tomatoes, herbs, and salt. Process just to blend, then add the eggs, one at a time, processing until smooth and creamy after each addition. Pour the filling into the prepared crust and bake in preheated oven for 1 hour, or until firm. Let stand in the oven for 30 minutes after baking.

Refrigerate until well chilled before slicing. Garnish each serving with a dollop of sour cream, a sprig of rosemary, and a small cluster of champagne grapes, if desired.

(Hilltop Café)

Hilltop Café

Fredericksburg

"**I**f you build it, they will come." Johnny and Brenda Nicholas must have understood that premise long before the popular movie *Field of Dreams* made it a national catchphrase. Back in 1981, when the pair bought the old gas station, grocery store, and beer joint on a lonely hilltop ten miles north of Fredericksburg, they were either very confident or very unaware that location means everything to the success of a restaurant. Whatever their reasoning, Hilltop Café has certainly been a success out there in the middle of absolutely nowhere. ❧ Some of that success may have to do with the fact that Johnny is an internationally known musician and a former member of the wildly popular western swing group Asleep at the Wheel. Or that he plays at the restaurant when he's not away on gigs with his own group, the Texas All Stars. Or it could be the wildly eclectic menu, featuring dishes from Johnny's Greek heritage and Brenda's Cajun roots, plus some Texas favorites like chicken-fried steak and grilled rib eyes priced by the ounce. Johnny credits the word-of-mouth reputation of the food for the success of the place. ❧ Johnny and Brenda try to balance the menu, offering something for everybody and in all price ranges. They use as many local ingredients as possible and concentrate on fresh, not frozen, products. "You won't find froufrou here," Johnny says, "just great down-home food." Hilltop is also known for its very unique wine list, which offers a wide spectrum of varietals and labels as well as many varietals by the glass. ❧ The Nicholases haven't changed the place: the antique gas pumps still stand sentinel in front, the interior is covered with memorabilia, the wood floors have a patina of wear, and the tables and chairs are an assortment of various styles. It's all part of the café's charm. ❧ A visit to Hilltop Café provides a scenic road trip with a group of adventurous dining friends on Friday and Saturday nights, but be sure to call for reservations at least a week in advance, or you'll find yourself inconveniently located ten miles from Fredericksburg with not the slightest chance of getting dinner! ❧ Hilltop serves one of the Hill Country's best brunches on Sundays, but expect a crowd.

10661 U.S. 87 North • (830) 997-8922 • www.hilltopcafe.com

House. Wine.

Fredericksburg

House. Wine., the brainchild of wine expert Todd Smajstrla, opened in the summer of 2007. It's a very unique concept, offering upscale home furnishings and accessories, well-selected antiques, and fabulous wines. The sleek, inviting space gives the illusion of walking into a very well-designed urban apartment or loft. But everything's for sale. The retail space flows seamlessly into the wine bar space, which is likewise inviting. The long bar offers comfortable seating, as do bistro tables spaced in front of the wine racks. For more intimate seating arrangements, there's the lounge, furnished with small groupings of leather couches and chairs—perfect for relaxing with a group of friends. ❖ House. Wine. offers an amazing selection of wines of myriad varietals. All wines are available by the glass or bottle, and Todd and his knowledgeable staff are happy to help you select a varietal you'll like. ❖ There's also a nice selection of wine-friendly tapas-style plates available. It's a great place to stop for a glass of wine at the end of the day or before going to dinner. House. Wine. has become a favorite with locals.

327 E. Main Street • (830) 997-2665 • www.intohousewine.com

Marburger Orchard

Fredericksburg

Marburger Orchard, just south of Fredericksburg, is a yearly destination for lots of folks from all over Texas. Come March, they start descending on the vast fields at Marburger to pick their own strawberries. You'll see whole families spaced out over two or three rows at a time. You can pick as many as you've got the fortitude to pick. ❧ After the strawberries have run their course in May, then there are big, succulent blackberries for the picking, followed closely by Fredericksburg peaches, the best in the country! ❧ Marburger Orchard, owned by Gary Marburger, has been a Hill Country tradition since 1978. It's one more farm dedicated to sustainable agriculture and the "eat local" movement. The fruit here isn't shipped across the country to be processed in a warehouse and then loaded onto a truck again to wind up in the supermarket! It goes from the orchard straight to the consumer. ❧ If you just don't feel like walking the fields, Marburger offers prepicked fruits for purchase. You can even taste the different varieties of peaches before you buy. ❧ Marburger provides the heavy cardboard containers for picking, so you can leave your odd assortment of buckets at home. Folks are even encouraged to save the containers from year to year—this saves a lot of trees!

550 Kuhlmann Road • (830) 997-9433 • www.marburgerorchard.com

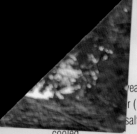

Yield: 16 to 18 kolaches

...east

...r (110–115 degrees)

...salted butter, melted and cooled

2 large eggs

1¼ cups sugar

2 teaspoons salt

8½ cups all-purpose flour

Peach Filling

4 large peeled and chopped Fredericksburg peaches (2 cups)

1 cup good-quality peach jam, melted

Streusel Topping

½ cup all-purpose flour

½ cup sugar

3 tablespoons chilled unsalted butter, cut into ½-inch cubes

Warm the milk in a saucepan over medium heat until it is steaming and beginning to form a skin on the surface. Do not boil. Remove from heat and cool 10 to 15 minutes, or until an instant-read thermometer registers 110–115 degrees. Dissolve the yeast in the lukewarm water and set aside until foamy, about 5 minutes.

In a large bowl, whisk together the eggs, sugar, salt, and melted butter. Add the cooled milk and yeast. Gradually add the flour to the batter, 2 cups at a time. Use your hands or a wooden spoon to mix the dough. Keep adding the flour until it is completely incorporated and the dough begins to hold together. Don't overwork the dough. It should be sticky, moist, and light.

Lightly grease a large bowl with vegetable oil. Transfer the dough to the bowl, cover with plastic wrap, and set aside to rise in a warm, draft-free spot until doubled in size, 1 to 2 hours. (When the dough has risen enough, a dent should remain when it is touched lightly.) Punch the dough down, cover with plastic wrap, and refrigerate overnight, or at least 4 hours.

(continued)

Rather Sweet Bakery and Café

Fredericksburg

Opened in 2001 by Rebecca Rather and Dan Kamp, Rather Sweet has become a morning gathering place with locals, who vie for tables with the hordes of visitors who line up outside the door every morning. ❧ Rebecca is a self-taught baker and the former pastry chef at Houston's renowned Tony's Restaurant. She creates some of the finest pastries, cakes, and upscale desserts in the country, many of them displayed in an antique bakery case. ❧ That pastry case holds dozens of items that have become local must-have breakfast goodies: Bacon-Cheddar Scones, Rebecca's plate-size Jailhouse Cinnamon Rolls, seasonal fruit kolaches, gigantic seasonal muffins, and more. Rather Sweet also makes one of the best breakfast tacos in town with house-made salsa to spice 'em up! Desserts consist of a too-tempting array of fruit tarts, lemon bars, Rebecca's signature Texas Big Hairs Lemon-Lime Meringue Tarts, a stunning variety of cakes (whole or by the slice), and many others. ❧ The bakery is a homey sort of place, located off Fredericksburg's busy Main Street in the rear of a grassy courtyard with a lovely old fountain. It feels somehow isolated from the hustle and bustle of shopping. You can dine outside when the weather's nice. The inside of the bakery offers seating in brightly colored spaces, creating an inviting ambiance. ❧ Rather Sweet serves both breakfast and lunch. Lunch features a delectable assortment of soups, salads, and sandwiches, plus a daily "hot lunch." All items are made from scratch in the bakery's kitchens. Whatever you order, be sure to save room for dessert!

249 E. Main Street • (830) 990-0498 • www.rathersweet.com

To make the filling, stir together the chopped peaches and melted jam; set aside.

Grease a 12-by-17-inch baking sheet with butter or cooking spray. With lightly oiled fingertips, shape the dough into balls about 2½ inches in diameter (the size of small limes). Arrange the balls evenly on the prepared baking sheet, 3 across and 6 down.

Using your thumb or finger, make a generous indentation in the center of each dough ball, being careful not to pierce the bottom of the dough. Mound about 1 heaping teaspoonful of the peach filling in the indentations. Cover the rolls with a clean towel and set aside to rise in a warm, draft-free spot until almost doubled in size, about 1 hour.

Preheat oven to 375. To make the Streusel Topping, place all ingredients in work bowl of food processor and pulse until crumbly. Scatter the topping over the kolaches just before baking. Bake in preheated oven for 25 to 30 minutes, or until lightly browned. Cool on wire racks for 20 minutes before serving. Serve warm or at room temperature.

(Rather Sweet Bakery and Café)

FREDERICKSBURG PEACH BREAD PUDDING WITH PEACH AND WHISKEY SAUCE AND CHANTILLY CREAM

Yield: 12 servings

8 ounces white chocolate, cut into small chunks

2 cups half-and-half

¼ pound (1 stick) unsalted butter, softened

½ teaspoon ground cinnamon

10 ounces day-old croissants, cut into ½-inch pieces

3 eggs

¾ cup sugar

7 medium fresh Fredericksburg peaches, peeled and chopped into bite-size pieces (about 3 cups)

½ cup toasted chopped pecans

1 tablespoons vanilla extract

Peach and Whiskey Sauce

½ pound (2 sticks) unsalted butter, softened

1½ cups sugar

2 eggs, beaten until frothy

⅓ cup peach schnapps

3 tablespoons good-quality sour mash whiskey

Chantilly Cream

1 cup whipping cream, well chilled

2 tablespoons sour cream

2 tablespoons powdered sugar

1 tablespoon vanilla extract

Preheat oven to 350 degrees. Butter a 13-by-9-inch baking dish; set aside. In a 2-quart saucepan combine the white chocolate, half-and-half, butter, and cinnamon. Cook over medium-low heat until smooth, stirring often. Remove from heat and set aside. Place croissant pieces in a large bowl and add the white chocolate mixture, blending well and breaking up the croissants. In bowl of electric mixer, combine the eggs and sugar; beat at medium speed until thickened, about 7 minutes. Add the peaches, pecans, and vanilla; beat just to blend. Fold the egg mixture into the croissant mixture, blending well. Turn out into prepared baking dish and bake in preheated oven for 45 to 55 minutes, or until a knife inserted in center comes out clean. Set aside and keep warm.

(continued)

Studebaker Farm

Fredericksburg

Russell Studebaker is an eighth-generation descendant of Studebakers who came to America from Solingen, Germany, in 1736 and settled in the Midwest. In the 1930s, Russ's paternal grandfather traded his farm in Missouri for land in the Rio Grande Valley. His maternal grandfather, a Swede, had moved to the region in 1913 and was one of the pioneers in developing Rio Grande Valley agriculture. Both farms remain prominent Valley farms, run by Studebaker cousins. ❦ Although Russ's father gave up farming during the long drought of the 1950s, he kept a small cattle herd and loved to plant fruit trees. So from an early age Russ knew the basics of growing fruit. ❦ In 1992 Russ and his wife, Lori, bought land east of Fredericksburg in the community of Blumenthal. Except for a twenty-five-acre orchard, the property was pretty much grown up in mesquite and cedar. The Studebakers cleared and leveled the land, planted cover crops to rebuild the soil, and built a home as well as a farm stand for selling the farm's peaches.

❦ Russ manages the orchard and the harvest, while Lori runs the retail end of the operation. They've not only expanded the orchard on their land but also leased other land, increasing their productive acreage to sixty acres with around 8,000 trees. ❦ The Studebakers enjoy the farming lifestyle, and their children have grown up working alongside Russ and Lori to make the farm a success. ❦ Studebaker Farm has repeat customers from all over the state who come here to buy strawberries, blackberries, peaches, and plums. Seasonal produce is also available. Many customers bring the Studebakers jars of preserves they've made from the farm's peaches. Kids come to see Rosebud, the family dog who goes to the fruit stand with Lori every day.

9405 U.S. 290 East • (830) 990-1109 • www.texaspeaches.com/studebaker

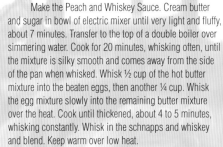

Make the Peach and Whiskey Sauce. Cream butter and sugar in bowl of electric mixer until very light and fluffy, about 7 minutes. Transfer to the top of a double boiler over simmering water. Cook for 20 minutes, whisking often, until the mixture is silky smooth and comes away from the side of the pan when whisked. Whisk ½ cup of the hot butter mixture into the beaten eggs, then another ¼ cup. Whisk the egg mixture slowly into the remaining butter mixture over the heat. Cook until thickened, about 4 to 5 minutes, whisking constantly. Whisk in the schnapps and whiskey and blend. Keep warm over low heat.

Make the Chantilly Cream by combining all ingredients in bowl of electric mixer. Beat at medium speed until the mixture forms loose, floppy peaks. Cover tightly with plastic wrap and refrigerate until ready to serve.

To serve, slice the warm pudding into squares. Ladle a portion of the Peach and Whiskey Sauce into each serving bowl and set a square of pudding in the center. Top with a large dollop of Chantilly Cream.

(Terry Thompson-Anderson)

LAVENDER STICKY BUNS

This really yummy recipe, created by Kimberly Brunner of Hannah's on Main in Fredericksburg, just had to be paired with a great lavender farm like Darlene Marwitz's Villa Texas Lavender Farm. These buns could be the favorite item at your next brunch. And you thought lavender was just for potpourri!

Yield: About 16 large buns

Dough Starter
2 cups cake flour
2 cups oat flour
3 cups water
⅛ teaspoon instant-rise yeast

Dough
Dough starter
3 cups all-purpose flour
½ cup nonfat dry milk
½ cup instant mashed potato flakes
½ cup sugar
4 teaspoons vanilla extract
1 tablespoon fine-grain sea salt
2 teaspoons instant-rise yeast
6 ounces (1½ sticks) butter, softened

Filling
2 cups sugar
3 tablespoons ground cinnamon
3½ tablespoons culinary lavender
6 tablespoons butter, softened

Glaze
1 cup corn syrup
2 tablespoons rum
6 tablespoons melted butter
2 cups light brown sugar

To make the dough starter, mix all ingredients in a nonreactive 5-quart bowl until well blended. Cover with plastic wrap and let sit on the counter overnight, but no longer than 16 hours.

(continued)

Villa Texas Lavender Farm

Fredericksburg

Darlene Marwitz is passionate about all things Italian. During travels to Italy as a graduate student in architecture, Darlene fell in love with Italian architecture and the sensuous landscape of rural Italy. She was also struck by the food and countryside of Tuscany and Umbria and its similarities to her own childhood memories of rural Texas, where farming and ranching were an inherited legacy. She was so inspired by Italy that she wrote a book, *Italy Fever: 14 Ways to Satisfy Your Love Affair with Italy*. ❧ Darlene has continued to express that appreciation for all things Italian. In the fall of 2003, Darlene and her husband, David, with the help of friends, planted an acre of lavender in a pastoral setting beside their home on the banks of the Pedernales River. "Whew!" says Darlene. "What a journey it has been. Too much rain. Too many weeds. Not enough sun early on—hard to believe in Texas, but true!" ❧ Today the lavender on Darlene's farm is lovely, some of the most beautiful in the Hill Country. You can visit the farm during the annual Fredericksburg Lavender Festival in early June. Cut your own blooming lavender sprigs and browse the broad spectrum of lavender products (including lavender plants), available at the rustic but elegant shop on the farm, where Darlene is available to share lavender lore.

4273 Morris Ranch Road • (830) 997-1068 • www.villatexas.com

Mix the dough by combining all ingredients in bowl of electric mixer fitted with dough hook. Mix at low speed until ingredients are moistened, then increase to medium speed and mix until dough is soft and smooth, about 10 minutes. Transfer to a large, lightly greased bowl, cover with plastic wrap, and refrigerate overnight. When ready to proceed with recipe, remove dough from refrigerator and set aside.

Make the filling by combining all ingredients except butter in work bowl of food processor fitted with steel blade; process for about 2 minutes. Turn out into a small bowl and set aside.

Butter two 9-by-13-by-3-inch sheet cake pans and line with parchment paper; set aside. Make the glaze. Combine the corn syrup, rum, and melted butter, whisking to blend well. Divide the mixture between the prepared pans. Scatter 1 cup of brown sugar evenly over the syrup in each pan.

Divide the dough into two equal pieces, rolling each out into a ¼-inch-thick rectangle. Spread 3 tablespoons of the softened butter over each rectangle, leaving a border of 1 inch around the edges. Sprinkle equal portions of the lavender filling over the butter and roll into logs, starting at the long sides of the rectangles. Slice the logs into 3-inch sections and place the buns, cut sides down, on the syrup in the cake pans. The buns should be fitted tightly against each other in the pan. Place in a warm spot, cover very loosely with a clean kitchen towel, and let rise until doubled in size, about 1 hour.

Preheat oven to 350 degrees. Bake buns in preheated oven for about 30 minutes, or until dough is golden brown and glaze is bubbling on the bottom. Let cool in pans on a wire rack for about 10 minutes, then invert onto separate baking sheets. Carefully remove parchment paper. Serve warm.

(Kimberly Brunner)

Vogel Orchard
Fredericksburg

The Vogel peach lineage goes back to the early 1900s when Armand Vogel planted a few peach trees and sold peaches and eggs under a shade tree to passersby on the road between Stonewall and Fredericksburg. ❦ Armand's son, George, and his wife, Nelda, planted their first orchard in 1953. From just 200 trees that orchard has now grown to over 2,800 trees. In 1972 the Vogels opened a roadside market ten miles east of Fredericksburg near the location where Armand used to sell under that shade tree. The Vogels' youngest son, Jamey, and his wife, Terri, moved back to Fredericksburg in 1998 to help run the family business. ❦ Vogel Orchard continues to offer much of the same produce that it has for many years—peaches, tomatoes, plums, blackberries, watermelons, cantaloupes, and other vegetables. In addition the roadside market sells homemade peach cobbler, peach ice cream, peach preserves, peach butter, plum jelly, and fig, blackberry, and pear preserves. ❦ Seventeen varieties of peaches are usually available, beginning around the middle of May with Starlite White clingstone peaches. The first freestone variety, Tex Royal, is usually ripe toward the end of May. Plums are harvested in mid-May. The Vogels grow Methley plums, a small purple-red plum that's sweet as honey. They're so small that at first the Vogels had trouble selling them, so they'd just hand out a few to folks. One taste and they were hooked!

12862 U.S. 290 East • (830) 644-2404 •
www.vogelorchard.com

Wildseed Farms

Fredericksburg

Wildseed Farms is a very unique place. Founded in 1983 by John and Marilyn Thomas, the farm has become a wildly popular destination for visitors to Fredericksburg from all over the world. ❊ The 200-acre farm grows wildflowers to produce seeds for over eighty varieties of wildflowers. John pioneered the practice of planting wildflowers in rows on large acreage, and today it's the nation's largest working wildflower farm. ❊ Fields of riotous color erupt at the farm when the wildflowers bloom from March through October. Visitors can explore walking trails through the blooming fields or pick their own wildflowers in specially designated areas. A popular attraction is the Butterfly Haus, a special exhibit of 250 species of live butterflies. ❊ The Market Center sells a vast array of wildflower seeds for all parts of the country in packets of individual species as well as mixes. The seeds are also sold through the Web site and a mail-order catalog. The Brew-Bonnet Biergarten, located in the Market Center, is a delightful place to enjoy a glass or bottle of Texas wine or beer with some vittles from the deli and to browse the vast selection of Texas food products available. ❊ Since 2007, Wildseed Farms has hosted an annual Gourmet Chili Pepper and Salsa Festival, which features seminars on all aspects of chili cultivation and chili culture. Dozens of varieties of chili plants are for sale, including some that are usually hard to find. ❊ Wildseed Farms is also a great source for garden plants—tomatoes, other vegetables, and herbs. You can find some of the more exotic herbs in addition to the familiar ones like dill and basil. Best of all, the helpful and knowledgeable staff is on hand to give pointers on how to best grow and care for the plants they sell.

425 Wildflower Hills (off U.S. 290) •
(800) 848-0078 •
www.wildseedfarms.com

GRAPE TOMATO PICO DE GALLO

This very different pico de gallo uses the chiltepin (or chilipitín), commonly known as the bird pepper, the only wild pepper in Texas. In fact, the chiltepin is officially designated as the Texas state native pepper. It's a tiny, volatile chili with a heat level of 8 on a scale of 10; however, the heat dissipates very quickly in the mouth and the resulting taste is quite nice. If you wish to make this condiment ahead of time, don't add the salt until ready to serve.

Yield: About 2½ cups

1 pint grape tomatoes, quartered
⅔ cup chopped cilantro
6 green onions and tops, sliced thin
6 chiltepins, minced, including seeds
Juice of 1 lime
Salt to taste

Combine all ingredients in a bowl and toss to blend well. Refrigerate in a tightly covered container until ready to serve. Serve with homemade tortilla chips or use as a condiment.

(Terry Thompson-Anderson)

CHOCOLATE CAYENNE CAKE

Yield: One 9-inch two-layer cake

2 cups sugar

½ cup cocoa powder

2 cups all-purpose flour

2 teaspoons baking soda

1 teaspoon baking powder

½ teaspoon red (cayenne) pepper

1 cup vegetable oil

1 cup buttermilk

1 cup water

2 teaspoons vanilla extract

2 eggs, beaten

Cream cheese frosting with anise (your favorite recipe with 1 teaspoon toasted and crushed anise seeds)

Chocolate Ganache

1½ cups bittersweet chocolate chips

3 tablespoons unsalted butter

2 tablespoons whipping cream

Preheat oven to 350 degrees. Generously grease two 9-inch cake pans and set aside.

Combine the sugar, cocoa, flour, baking soda, baking powder, and cayenne, stirring to blend well. In a separate bowl, combine the vegetable oil, buttermilk, water, vanilla extract, and beaten eggs. Add the wet ingredients to the dry ingredients, beating to blend well. Divide the batter between the prepared cake pans; bake in preheated oven for 30 minutes, or until a toothpick inserted in the center of the cakes comes out clean. Remove from oven and cool in pans on wire racks for 10 minutes. Turn the layers out onto the rack, allow to cool completely, and refrigerate until chilled.

To make the Chocolate Ganache, melt the chocolate chips and butter in a double boiler over simmering water; stir until smooth. Add the cream slowly, stirring until smooth and spreadable.

To assemble the cake, place one layer on a serving plate and spread with the cream cheese frosting all the way to the edge. Place the second layer on top of the icing. Frost the top and sides of the cake with the Chocolate Ganache. Allow the ganache to set before slicing.

(Elaine's Table)

Elaine's Table

Hunt

The charming limestone building that houses Elaine's Table was built in the late 1930s as a restaurant catering to summer residents, the families of campers, and hunters in the winter months. Although it's undergone a number of renovations, it's been used as a restaurant for most of its history. ❀ The present incarnation of the space offers an inviting Hill Country ambiance, with its limestone interior walls and Mexican tile floors. There's a fabulous view of the gracefully flowing Guadalupe River from the back dining area. The large fireplace in the main dining room offers a cheery glow during the winter months. ❀ Elaine Bicknell opened Elaine's Table in 1998, and soon her husband, Bruce, joined Elaine in operating the restaurant. The community support was evident from the beginning, and the restaurant quickly became a favorite meeting place for local residents, especially on Wednesday nights, when all the tables are grouped together for Family Night. Huge platters of all manner of goodies are placed in the center of the tables, family-style. ❀ The regular menu offers a variety of appetizers, soups, salads, and

entrées from seafood to steaks, pork, pasta, chicken, and a very good chicken-fried steak made with sirloin. The dessert table offers a tempting array of cobblers, cakes, pies, and crème brûlée. ❀ During the summer the restaurant is packed on the drop-off and pick-up weekends for the large summer camp industry in the area. Elaine's is also the "launching spot" (and ending celebratory point) for the annual Hunt Sweet Potato Queens Parade in February, a private event. The Bicknells say that their favorite thing about the restaurant is their great customers.

1621 State Hwy. 39 • (830) 238-4484

• www.elainestable.com

Broken Arrow Ranch

Ingram

Broken Arrow Ranch, suppliers of free-range, field-harvested wild game, was founded in 1983 by Mike Hughes and his wife, Elizabeth. On a vacation abroad the Hugheses had noticed that venison and other wild game meats were common on menus in European restaurants but almost nonexistent in American restaurants. At the same time, exotic deer species imported by Texas ranchers for hunting purposes were overgrazing Hill Country ranchlands, requiring active population management to maintain the sustainability of both the herd and the land. The Hugheses established Broken Arrow Ranch to provide high-quality wild game meats to the restaurant industry. ❄ Broken Arrow Ranch is the nation's only supplier of truly wild venison. The exotic species harvested by Broken Arrow range free on over 100 Texas ranches that total about one million acres. This land provides native vegetation for browse and gives the meat the complex natural flavors not found in that of farm-raised animals. The animals are range-harvested with techniques that reduce stress to the animal and produce the highest meat quality possible. Working closely with government inspection agencies, Mike developed a portable, trailer-mounted processing facility so that an animal can be processed quickly after being harvested. ❄ In 2005 Mike turned the reins over to his son Chris, who had already learned the business by doing product demonstrations at trade shows, hosting chef retreats at the family's ranch in Ingram, and taking physical inventory counts. ❄ The majority of Broken Arrow's products are sold directly to chefs and restaurants nationwide, but consumers can purchase cuts of venison, antelope, and wild boar for home use through the company's Web site. Eating wild game makes sense with today's health concerns. Broken Arrow's venison and antelope meat averages one-third the calories of beef, one-eighth the fat content of beef, and is lower in cholesterol than a skinless chicken breast!

3296 Junction Hwy. (State Hwy. 27) • (800) 962-4263 • www.brokenarrowranch.com

SAVORY VENISON STEW

The cooks at the ranch perfected this recipe in the course of feeding the annual hunting camp crowd.

Yield: 8 servings

2 pounds Broken Arrow Ranch venison stew meat, cut into bite-size pieces

2 bay leaves

1½ tablespoons mixed pickling spices, securely wrapped in cheesecloth and tied with kitchen twine

2 medium onions, coarsely chopped (2 cups)

¼ teaspoon granulated garlic

½ teaspoon ground black pepper

4 carrots, cut into ½-inch dice

4 celery stalks, cut into ½-inch dice

4 red potatoes, cut into ½-inch dice

4 chicken bouillon cubes

1 quart water

2 teaspoons red wine vinegar

2 cans (8-ounce) tomato sauce

Spray a heavy Dutch oven with nonstick vegetable spray and quickly brown the meat on all sides. Add all remaining ingredients, cover, and simmer over medium heat for about 1½ hours, or until venison is tender and vegetables are cooked. Remove and discard the cheesecloth bag before serving.

(Broken Arrow Ranch)

Friendly Bar Bistro

Johnson City

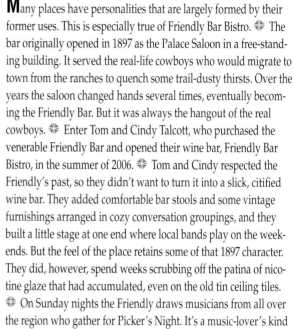

Many places have personalities that are largely formed by their former uses. This is especially true of Friendly Bar Bistro. ❧ The bar originally opened in 1897 as the Palace Saloon in a free-standing building. It served the real-life cowboys who would migrate to town from the ranches to quench some trail-dusty thirsts. Over the years the saloon changed hands several times, eventually becoming the Friendly Bar. But it was always the hangout of the real cowboys. ❧ Enter Tom and Cindy Talcott, who purchased the venerable Friendly Bar and opened their wine bar, Friendly Bar Bistro, in the summer of 2006. ❧ Tom and Cindy respected the Friendly's past, so they didn't want to turn it into a slick, citified wine bar. They added comfortable bar stools and some vintage furnishings arranged in cozy conversation groupings, and they built a little stage at one end where local bands play on the weekends. But the feel of the place retains some of that 1897 character. They did, however, spend weeks scrubbing off the patina of nicotine glaze that had accumulated, even on the old tin ceiling tiles. ❧ On Sunday nights the Friendly draws musicians from all over the region who gather for Picker's Night. It's a music-lover's kind of evening when some two dozen or so pickers assemble out on the back deck and just play their hearts out. ❧ There's a nifty selection of wines by the bottle and glass, all well priced. And—hip hip hooray!—a good selection of Texas wines. Of course, as a nod to the former life of the place, there's a great list of beers, too, including the award-winning brews from Real Ale Brewing Company down the road in Blanco. ❧ After discovering that folks would come in shortly after opening time and stick around for a while, Tom added a little tapas-style menu of tasty bar food prepared by a local chef. ❧ Friendly Bar Bistro is a unique place. You'll still see the occasional cowboy bellied up to the bar with a longneck, but you're just as likely to see a cowboy at that bar with a balloon stem of Cabernet Sauvignon and a cowgirl with a longneck!

106 North Nugent • (830) 868-2182 • www.friendlybarbistro.com

Hummingbird Farms

Johnson City

Hummingbird Farms, nestled in the serene countryside west of Johnson City, is owned and operated by Jack and Debi Williams. The Williamses planted their first lavender field at Hummingbird Farms in 2003 after they had spent several years researching which types of lavender would be best for the optimal production of essential oil. The Williamses settled on *Lavendula angustifolia*, known variously as common lavender, English lavender, or true lavender. ❧ Hummingbird Farms produces high-quality bath and body products, using only lavender oil, aloe, and other pure ingredients grown in Texas, with no added water or artificial fragrances. The Williamses developed their own formulas for the products, which are available at gift shops and other specialty stores throughout Central Texas or via the Hummingbird Farms Web site. ❧ Hummingbird Farms is open to visitors during the lavender season from the end of May to the end of June and offers cut-your-own lavender.

9430 U.S. 290 West • (830) 868-7862 • www.hummingbirdlavender.com

HILL COUNTRY PEACH SALAD
WITH PEACH-LAVENDER VINAIGRETTE

Yield: 6 servings as an entrée salad,
8 servings as a side salad

16 cups seasonal mixed greens
1½ cups crumbled gorgonzola cheese
¾ cup toasted pecans
6 peaches, peeled and sliced

Peach-Lavender Vinaigrette

1 cup rice wine vinegar
2 teaspoons culinary lavender flowers
2 tablespoons Dijon-style mustard
2 teaspoons salt
2 teaspoons black pepper
1 cup peach preserves
2 cups canola oil

Make the Peach-Lavender Vinaigrette. Combine all ingredients except the peach preserves and canola oil in work bowl of food processor fitted with steel blade; process until smooth and well blended. Stop the processor and add the peach preserves. With processor running, add the canola oil in a slow, steady stream through the feed tube. Process until only small pieces of peaches from the preserves remain.

Place seasonal greens in serving bowl, sprinkle with gorgonzola, and add the pecans and peaches. Toss the salad with Peach-Lavender Vinaigrette and serve at once. The remaining vinaigrette can be stored in the refrigerator for up to a month.

(Silver K Café)

Silver K Café

Johnson City

If you want a great meal, try to pass through Johnson City at either lunchtime or dinnertime so you can stop at the Silver K Café. The place is a living example of the best of Texas-friendly service and traditional Hill Country cuisine in an atmosphere of laid-back, rustic elegance. ❄ The Silver K was established by Kay Pratt and Karen Granitz in 2002 in a storage barn of the historic Old Lumber Yard (formerly the Stein Lumber Company). During the remodeling process they tried to keep as much of the old wood as possible. ❄ Today Kay and her husband, Al, carry on the vision at the café. It has been a success story since the beginning, having been featured in numerous publications, including *Gourmet*, *Texas Monthly*, *Texas Highways*, and many more. ❄ From Sunday through Wednesday the menu features "Homestyle Suppers," including such comfort food as meat loaf and chicken pot pie. From Thursday through Saturday the menu changes gear to upscale dining, offering a selection of steaks (all corn-fed, dry-aged beef), seafood, poultry, and pasta. The chicken-fried steak, served with country gravy, is one of the best in the Hill Country. The Silver K Café proudly features an extensive selection of Texas wines on its wine list. There's a fabulous Sunday buffet, and one of the most extensive breakfast menus you'll find anywhere. It's loaded with everything you could possibly dream of wanting for breakfast—from omelets to migas and everything in between. ❄

The Pratts' love of music led them to add Estrella, a special all-weather outdoor area where folks can enjoy their great food along with great music from Texas singers and songwriters. ❄ The Silver K Café used to be a hidden Hill Country find, but now it's been "found," so get there early!

209 E. Main Street • (830) 868-2911 • www.silverk-cafe.com

Texas Hills Vineyard

Johnson City

Following in the footsteps of many other entrepreneurs who have given up lucrative professions to move to the Texas Hill Country and "do something different," pharmacists Gary and Kathy Gilstrap sold their successful drugstore in Kansas and headed to the Johnson City area. They were looking for just the right vineyard land, which they found east of town on the road to Pedernales Falls State Park. ❧ The Gilstraps first became interested in wine while traveling in Europe. They began to study wine production methods, applying their knowledge of chemistry and biochemistry. Once they found their "dirt," Kathy's son, Dale Rassett, decided to join the venture and now manages the vineyard. The trio advocates "sustainable organic" growing techniques, which involve using the least amount of chemicals possible. Turkey compost is used to fertilize the vineyard, and weeds are removed with a mechanical device, made especially for organic vineyard cultivation, rather than herbicides. ❧ The Texas Hills team is also dedicated to looking out for the Hill Country environment. The winery and tasting room facility was built using the rammed-earth method to cut down on the amount of energy needed to cool the tasting room—and to make it a very quiet and cozy space. ❧ The land reminded the Gilstraps of the Tuscan countryside, so they initially decided to grow Italian varietals, from which they produce some very noteworthy, award-winning wines. The first year of production was 1996, and the vineyard has grown from twelve acres of vines to thirty-five acres. They were the first in Texas to attempt to grow pinot grigio, and Texas Hills Vineyard's Pinot Grigio is a fine, food-friendly wine. In addition, they grow sangiovese, cabernet sauvignon, merlot, syrah, and moscato. ❧ The winery has made a name for itself on restaurant menus with several of its varietals and two great dessert wines, Moscato and Orange Moscato, plus a ruby port–style wine named Rubino—fabulous with anything chocolate or just for sipping. Texas Hills Vineyard's premium Cabernet Sauvignon, Kick Butt Cab, is two-fisted and full-bodied, a must-have accompaniment to those cowboy rib eyes that we love so dearly in the Hill Country!

878 RR 2766 • (830) 868-2321 • www.texashillsvineyard.com

89

BUTTERNUT SQUASH SOUP

Angel Valley Organic Farm grows a variety of winter squashes, including delicious butternut squash. This squash soup is comfort food to the max. Fill a tall mug with the soup and curl up in front of a warm fire for some quality "down time."

Yield: 6 to 8 servings

¼ pound (1 stick) unsalted butter

1 onion, chopped

1 large baking potato (about 12 ounces), peeled and sliced

1 sweet potato (about 10–12 ounces), peeled and sliced

3 cups peeled and diced butternut squash

2 jalapeños, seeds and veins removed, minced

1¼ teaspoons minced fresh ginger

½ cup all-purpose flour

1½ quarts chicken stock

2 teaspoons real maple syrup

Salt to taste

¼ teaspoon red (cayenne) pepper, or to taste

1 cup whipping cream

Sour cream

Chopped toasted pecans as garnish

Melt the butter in a heavy 6-quart soup pot over medium heat. Add the onion and sauté until wilted and transparent, about 5 to 6 minutes. Add the potato, sweet potato, squash, jalapeños, and ginger. Toss to coat with the butter. Add the flour all at once and stir to blend well. Cook over medium heat for 3 to 4 minutes, stirring, until all flour is blended into the butter. Add the chicken stock and maple syrup. Season to taste with salt and cayenne. Bring to a boil to thicken the soup, stirring often. Cover the pan, lower heat to a simmer, and cook until all vegetables are very tender, about 30 to 45 minutes. Puree the soup in batches in a food processor or blender until very smooth. Return the soup to a clean pot and stir in the whipping cream. Cook just until the cream is heated.

To serve, ladle the soup into shallow soup plates; garnish with a dollop of sour cream and scatter toasted pecans over the sour cream.

(Terry Thompson-Anderson)

Angel Valley Organic Farm

Jonestown

Farming was never in the plan that John and Jo Dwyer set out for their lives. Jo, who grew up in Indiana's "corn country," thought she'd had her fill of flat farmland, and John was strictly a product of suburban America. It wasn't until they relocated to Austin in 1979 that he was bitten by the gardening bug. He leased a plot in the community gardens and learned how to grow vegetables organically. ❋ With the purchase of their first home in Jonestown, John planted the garden of his dreams, complete with a wooden greenhouse constructed from scratch. Through the years the Dwyers climbed the corporate ladder, she as a paralegal, he as a co-owner in an auto body parts company. When John and his partners sold the business, he approached Jo with the idea of starting a farm. They purchased fifteen acres in a sheltered valley in Jonestown in 1996, and Angel Valley Farm became a reality. ❋ They erected a deer-proof fence, brought in 50 tons of manure for the first three acres to be planted, and added cover crops to further enrich the ranchland soil. They sold their first crop at a farmer's market in Austin in the spring of 1999. It was a good beginning. Now the Dwyers do rotational planting on eight acres, using only about two acres at a time. The farm has been certified organic by the Texas Department of Agriculture from the very beginning. ❋ Angel Valley's produce is serious produce—and beautifully clean. On a summer day you'll

find such treasures as tomatoes (including cherry tomatoes and delicious heirlooms), cucumbers, Red Lasoda potatoes, 1015 yellow onions, elephant garlic, bags of tender young arugula and bunches of basil, green and white bell peppers, cubanelle peppers, three varieties of eggplant, okra, squashes, and whatever melons are ready for harvest (including lots of crispy Asian melons). Cooler weather brings a bounty of fall crops like beets, broccoli, cabbage, carrots, cauliflower, leeks, onions, peas, potatoes, turnips, spinach, winter squashes, and several varieties of greens. ❋ Angel Valley Organic Farm products are available at two farm stands that the Dwyers operate themselves: at the Asian American Cultural Center on Jollyville Road in Northwest Austin (Wednesdays) and at the Plaza Center on FM 1431 in Jonestown (Saturdays).

18405 Angel Valley Drive • (512) 267-2785

• www.angelvalleyfarms.com

Milky Way Drive-In

Junction

As back-roads travelers know, independently owned drive-ins often have some pretty outstanding vittles, and this Junction landmark is no exception. ❧ The Milky Way was built in the early 1950s by a World War II veteran who had returned home and wanted to improve the quality of life in his town. He was also the co-owner of Junction's first drive-in movie theater, the Moon-Glo. The Milky Way was originally a small stand that sold soft ice cream, hence the name. Later the menu was expanded to include hamburgers. The Milky Way passed through many hands over the years but remained a popular place. ❧ In 2005 Californians Catherine and Marty Dolfuss bought the Milky Way. Junction had become a favorite getaway for the couple during the hunting season, and

they always grabbed a burger from the Milky Way when they came to town. When they discovered that the place was for sale, they bought it and moved to Junction. ❧ The new owners have spruced up the place, adding tile to the dining room and rest rooms, a real pizza oven, and new coolers and freezers. And they expanded the menu to include those handmade pizzas, Mexican foods, fried catfish, an awesome hand-pounded chicken-fried steak, and the best Patty Melt I've ever had. ❧ Over the years the Milky Way had become known for excellent milk shakes, so the Dolfusses didn't mess with those. Another favorite, the burgers, are pretty much the same but made with better

meat. Everything is made from scratch—a rarity at a drive-in, including hand-cut fries. All meats and the catfish are breaded when they're ordered, and the burger patties are shaped by hand! The drive-in, which is open for both lunch and dinner, even makes its own tasty salsa for the tacos! ❧ Slow fast food—you gotta love it!

1619 Main Street • (325) 446-2695

PECAN CRESCENTS

Yield: 16 crescents

2 cups sifted all-purpose flour
¼ teaspoon salt
½ pound (2 sticks) frozen unsalted butter,
    cut into 1-inch cubes
1 egg yolk, beaten
¾ cup sour cream

Pecan Filling

¾ cup sugar
1½ teaspoons ground cinnamon
1 cup coarsely chopped pecans

In food processor fitted with steel blade, combine the flour and salt. Add the butter, pulsing to break it into pea-size chunks. In a small bowl combine the egg yolk and sour cream, whisking to blend; add to flour mixture. Process just to combine. On a lightly floured surface, knead the dough gently three or four times, or just enough to form a cohesive dough. Divide into 4 equal portions and pat each portion into a disk. Wrap each disk in plastic wrap; refrigerate overnight.

Prepare the Pecan Filling by combining all ingredients in a medium bowl and tossing to blend; set aside. Preheat oven to 375 degrees. Remove the dough from the refrigerator, 1 portion at a time. On a lightly floured surface, roll out dough into a 12-inch circle. Cut into 8 wedges using a sharp knife or pastry cutter. Scatter one-fourth of the Pecan Filling evenly over the wedges. Roll up each wedge, staring at the outside edge. Place the wedges on an ungreased baking sheet and bend into crescent shapes, with the curving ends pointing in toward each other. Repeat with the remaining disks. Bake in preheated oven until golden brown and crispy, about 20 to 25 minutes. Cool on a rack.

(Terry Thompson-Anderson)

South Llano Farms

Junction

The rich bottomland soils of the South Llano River make the Junction area prime pecan-growing country. South Llano Farms consists of some 250 acres of riverfront property with native pecan trees planted by Mother Nature. Some of the trees are hundreds of years old. Irrigated by the spring-fed Llano River, the trees produce 15,000 to 20,000 pounds of pecans each year. Reginald and Becky Stapper, who had always yearned for the day when they could have a place of their own, bought the property in 1997 as a joint venture with John and Martha Watts. They also leased an additional 80 acres, part of the old 7C Pecan Orchard, planted in the 1950s. The harvest from this orchard, planted in Wichita and Western pecan varieties, averages 40,000 pounds a year. South Llano Farms has a storefront operation for processing and shipping. The store, open seasonally, also sells fresh pecans by the pound and a variety of pecan-studded goodies made from recipes developed by John and Martha, along with their family and friends—scrumptious pecan pies to order, cinnamon pecans, roasted pecans, pecan clusters, and pecan cookies. The farm's products and locally made jam, jellies, salsas, sauces, and assorted gift baskets are available year round via the Web site. South Llano Farms is a small operation, a family farm trying to compete with large producers and raise a quality product. The Stappers work the farm, endeavoring to make it profitable, while raising their four daughters. They also grow coastal hay and raise sheep and goats. With a little divine providence, the weather will be kind, and they'll survive the August threat of pecan weevils and have successful harvests in late October and November.

550 Old Highway 377 • (325) 446-8271 • www. southllanofarms.com

Rails—A Café at the Depot

Kerrville

Does the concept of healthy food that tastes really good sound like an oxymoron? Rails has proven that it can be done. ❦ Owners Melissa Southern and John Hagerla opened the popular café in 2003. Melissa, a registered dietician, was tired of unappealing, tasteless hospital food. She wanted to open a restaurant serving healthy fare that was also delicious. ❦ The café is located in the old depot of the San Antonio & Aransas Pass Railway. Built in 1915, the depot was used until 1970, when the railroad line was abandoned. In renovating the building, Melissa and John created a friendly, laid-back kind of place, not stuffy and

upscale. The interior is accessorized with an interesting collection of old railroad memorabilia. The restoration was named Best Rehabilitation Project in 2004 by the Texas Downtown Association.
❦ Melissa's menu is a delight, offering something for the customer's every mood. There's a large selection of appetizers, some great sandwiches (including an Axis Deer Burger and Grilled Italian Panini), and a variety of salads, full entrées, and daily dinner specials, including a great Pork Shank Osso Buco. Open for lunch and dinner, Rails has a nice wine list, with most of the selections offered by the glass. ❦ Rails has won accolades from *Frommer's San Antonio and Austin* ("Some of the best food in the Hill Country") and the *San Antonio Express-News* ("A small gem polished with local flavor and a bit of history"). And remember, it's all healthy!

615 E. Schreiner Street • (830) 257-3877 • www.railscafe.com

SEARED DUCK BREAST WITH HILL COUNTRY PEACH AND GINGER SAUCE

Yield: 4 servings

- 4 tablespoons olive oil
- 1 cup chopped shallots
- ½ cup sugar
- ¾ cup low-sodium soy sauce
- 1½ cups dry red wine
- ¼ cup balsamic vinegar
- 2½ tablespoons grated fresh ginger
- 1½ teaspoons ground cinnamon
- ½ teaspoon freshly ground black pepper
- 4 skinless, boneless duck breast halves
- 3 fresh Hill Country peaches, peeled and chopped

In a heavy 12-inch skillet heat 2 tablespoons of the olive oil over medium heat. Add the shallots and sugar, sautéing until shallots are golden brown. Add the soy sauce, red wine, balsamic vinegar, ginger, cinnamon, and black pepper, stirring to blend well. Cook over low heat for 2 minutes. Cool and reserve 1 cup of the sauce for marinade. Place the duck breasts in a single layer in a Pyrex baking dish. Pour the reserved 1 cup of sauce over the breasts, cover, and refrigerate for 2 hours. Turn occasionally.

Remove duck from marinade and pat dry with absorbent paper towels. Coat the duck breasts with olive oil and season with salt. Heat the remaining 2 tablespoons of oil in a heavy 12-inch skillet over medium-high heat until the oil begins to smoke. Add the duck breast halves and cook 3 minutes on each side, or until golden brown. Remove to a platter and keep warm.

In a heavy medium saucepan, boil the remaining sauce until reduced by half, about 5 minutes. Add peaches and stir until heated through. Pour the sauce over the duck breasts and serve at once.

(Rails—A Café at the Depot)

CHOCOLATE AND GOAT CHEESE CRÈME BRÛLÉE

The plain soft chèvre made by Bradley Goat Farms is outstanding in this recipe, which combines the tangy, earthy goat cheese with the best features of chocolate—for an over-the-top delicious dessert. Even goat-cheese haters love it.

Yield: Six 6-ounce ramekins

⅓ cup light brown sugar
2¼ cups whipping cream
4½ tablespoons sugar
3 ounces bittersweet chocolate
2 tablespoons Dutch-process cocoa
4 egg yolks
2½ teaspoons vanilla extract
7½ ounces goat cheese at room temperature

Preheat oven to 250 degrees. Spread the brown sugar in a thin layer on a baking sheet and place in preheated oven for about 10 minutes. Do not allow the sugar to melt. Remove baking sheet to cooling rack while preparing the custard. Increase oven temperature to 325 degrees. Lightly butter bottom and sides of six (6-ounce) ramekins; set aside.

Combine the whipping cream, sugar, chocolate, and cocoa in a heavy 2-quart saucepan over medium heat. Whisking often, cook until chocolate is melted and sugar has dissolved. Remove from heat and set aside to cool.

Combine the egg yolks, vanilla extract, and softened goat cheese in work bowl of food processor fitted with steel blade. Process until smooth and well blended. Turn the egg mixture out into a medium bowl and whisk in the chocolate mixture until smooth. Divide the mixture among the prepared ramekins and place the ramekins in a baking dish. Add enough hot water to reach halfway up the sides of the ramekins. Bake in preheated oven for 1 hour. Remove ramekins from baking dish and refrigerate until the custard is thoroughly chilled, at least 6 hours.

Put the cooled brown sugar in work bowl of food processor fitted with steel blade or an electric coffee grinder and process until the sugar is very fine. Set aside.

When the custard is well chilled, place the ramekins on a baking sheet and divide the brown sugar among the ramekins, spreading a thin layer on top of each one. Using a culinary blowtorch, caramelize the sugar, taking care not to burn it. (Or if you don't own such a gadget, preheat the broiler and place the oven rack 3 inches under the heat source. Place the baking sheet under the broiler and cook just until the sugar has caramelized.) Allow the sugar to harden for a few minutes, and then serve.

(Terry Thompson-Anderson)

Bradley Goat Farms

Lampasas

Lou Bradley is an authentic Texas treasure. A fourth-generation Texan, she and her husband moved to the Hill Country after Dave's retirement from the air force ended their prolonged odyssey of travel. The Bradleys have been in Lampasas for thirty-one years now, and Lou's been making goat cheese for twenty-five of those years. ❖ Soon after they had purchased the fifteen acres adjacent to their original seven-acre property, the Lampasas school district decided to raise taxes—a lot. Several hundred protestors attended the next school board meeting, and right in the middle of a rowdy group on the front row was Lou Bradley. Finally, one of the school board members leaned over to her and said, "Lou, get goats, declare the land for agriculture, sit down, and shut up!" She went home and told her husband that they getting some goats. She had no earthly idea what she would do with goats, but she wasn't about to pay those taxes! Not long ago Dave observed that maybe they should have paid those taxes instead of working so hard in their "retirement." ❖

What to do with a herd of Nubian and Alpine goats? Lou decided to open a goat's milk dairy and taught herself how to make cheese by reading and experimenting. Her first milk inspector encouraged her to get a license to manufacture cheese after tasting some of her trial batches. Boy, did she get it right—her goat cheese is some of the best I've ever had. But it isn't available in stores. You can buy Bradley Goat Farms cheeses only at the farm. Give Lou a call to be sure she has a good supply, and tell her when you'd like to come. ❖ Lou makes two types of cheese: soft chèvre and semi-hard. The mind literally races with the possible uses for her four flavors of soft chèvre: Plain, Green Chile, Chipotle, and Sun-Dried Tomato & Garlic! And her semi-hard cheese has a delicate earthy taste and a firm, almost crumbly texture; soaked in Cabernet Sauvignon, it's the perfect foil for the deep berry nuances of the wine. Don't need to think about a use for that one—just get out the cheese knife! Make the trip to Lampasas, and be sure to take along a big cooler.

1046 BCR 111 • (512) 556-3109

Eve's Café

Lampasas

Eve's Café is located in a charming building right on the square in downtown Lampasas, a picturesque Hill Country town in which life goes on much the way it always has—at a slower pace than the rest of the world. But there's nothing slow about the pace in the kitchen at Eve's. On Friday nights the place is standing-room-only. ❀ Eve Sanchez was born and raised in the small German town of Fulda, just north of Frankfurt. She married an American GI stationed in Germany, and her husband's eventual transfer to Fort Hood in Killeen brought Eve to the United States and its myriad opportunities. ❀ She began to muse on the fact that she had spent her whole life working for other people—and working very hard. So why not take that same capacity for hard work and put it to use in a place of her own? When her husband retired, the Sanchezes moved to Lampasas, where Eve discovered the wealth of German heritage in the surrounding small towns of the Hill Country. ❀ In 1995 Eve and her husband, Steve, opened Eve's Café, which quickly became a favorite dining spot in Lampasas, serving breakfast and lunch from Monday through Thursday and also dinner on Fridays.

The Sanchezes' son, Marc, has stepped in to run the place, but you'll still see Eve there most of the time when the restaurant is open. ❀ At Eve's you won't find sausage plates drowned in limp red cabbage or bland sauerkraut with a side of watery potatoes. What you will find is food from the small German villages, prepared in the traditional manner. Eve's specialty is pork schnitzel. In fact, Eve's serves just the kind of food that probably sustained the early German pioneers in the area. Eve's is a special treat right in the heart of the Texas Hill Country!

521 E. 3rd Street • (512) 556-3500

HOUSE SALAD

When sharing this recipe, Eve said, "You must have the German white vinegar! It won't work with other kinds of vinegar." And, of course, she was right. I tried it with several kinds of vinegar readily available at the supermarket, and it truly wasn't the same. So head for the specialty markets and scout out authentic imported German vinegar. Prepare the dressing at least 2 hours before serving and chill it well.

Yield: 6 servings

1 head romaine lettuce, cut into slices 1½ inches thick
Sliced cucumber
Thin-sliced red onion

House Dressing
2 green onions, minced
3 fresh dill sprigs, minced
1½ cups whipping cream
½ cup whole milk
3 tablespoons German white vinegar
1 tablespoon sugar
1 teaspoon salt

Make the House Dressing. Combine the green onions and dill in a medium bowl. Whisk in the cream and milk, blending well, then slowly whisk in the vinegar to form a smooth mixture. Whisk in the sugar and salt. If the dressing is too thick, add additional milk, a little at a time, whisking to blend. Adjust seasoning as needed. Cover and refrigerate until well chilled, or at least 2 hours.

To assemble the salad, place a portion of the romaine on each salad plate and drizzle a portion of dressing over the top. Arrange 3 or 4 cucumber slices on each salad and scatter a few of the thin-sliced red onions on top.

(Eve's Café)

GRANNY JOY'S MEAT LOAF

This family recipe for meat loaf was passed down to Sandra from her mother. Sandra says to serve it with a side of mashed potatoes and bread for sopping up the good gravy.

Yield: 8 to 10 servings

Meat Mixture

5 pounds lean ground beef
1½ cups chopped green bell pepper
1 cup chopped onion
2 cups finely ground seasoned croutons
1 cup grated Parmesan cheese
1 tablespoon kosher salt
1 tablespoon black pepper
4 eggs, well beaten
2 cups pureed diced tomatoes
6 (⅜-inch thick) slices of Velveeta cheese cut from a block
2 cups ketchup
1 cup crumbled, crisp-cooked bacon

Tomato Gravy

½ cup bacon drippings
½ cup diced onion
½ cup all-purpose flour
1 quart chicken broth
1 (15-ounce) can diced tomatoes and their juice
Salt and freshly ground black pepper to taste

Preheat oven to 325 degrees. In a large mixing bowl combine all ingredients for the meat mixture except cheese slices, ketchup, and bacon; mix well with the hands.

Divide the meat mixture in half. Pat half of it into

(continued)

Yumm Factory Café

Lampasas

Ever wish you could go back to your grandmother's kitchen for just one more meal? Well, they certainly won't look like your grandmother, but Sandra Jullian and Norma Spinner serve up a whole menu full of downright yummy home-style vittles at their Yumm Factory Café. ❧ It's a busy place that's jam-packed with a crowd of local regulars. The breakfast menu offers serious grandma breakfasts, including one with a New York strip steak! Lunch and dinner feature down-home favorites as well as a good

selection of steaks, salads, sandwiches, and really decadent seasonal fruit cobblers. For much more food than you can eat, try the daily special! ❧ Sandra grew up cooking for the farmhands at her family's farm in New Mexico. When she left home, she started a catering company in Fort Worth. Norma, a native Californian from a ranching family, came to work for Sandra's company. Then it became a family relationship when Norma's sister married Sandra's son! ❧ As a Christmas present in 2005, Sandra's husband, Ron, now deceased, bought the property on South Key to make his wife's lifelong dream of having her own restaurant come true. The Jullians recycled materials from the little house on the property for use in building the café. The tin ceilings came from the old courthouse in Lampasas. Other parts came from an old sheep barn in Adamsville and the roof of another old barn. The bar stools at the main counter came from an old drug store in Austin, and Sandra, her father, and Ron made all of the tables. ❧ The café opened in 2006 with almost 100 seats—half of them inside and half outside. Sandra never expected the overwhelming success of the place. Some customers eat there two or three times a week, and most of the menus are scribbled up with commentaries on the great food. Sandra laughingly says, "This isn't my restaurant. It belongs

to the people of Lampasas!" Sandra feels that the best thing about the restaurant is the interaction with the community—and meeting tourists who stop in for a bite to eat. Many come back the next time they pass this way.

1902 S. Key Avenue (U.S. 183/190) • (512) 556-0550 • www.theyummfactory.com

the bottom of a deep baking pan (such as a small roaster) with a lid. Be sure the meat goes all the way to the edges of the pan. Arrange the cheese slices over the meat, leaving a border of about one inch at the edges. Top the cheese with the remaining half of the meat mixture. Seal the edges of the meat mixture so the cheese doesn't leak out. Pour the ketchup evenly over the top of the meat loaf, cover, and bake for 2 hours in preheated oven. Check frequently toward the end of the cooking time. When the meat loaf is nearly done, remove the lid and scatter the crumbled bacon over the top. Cook, uncovered, until an instant-read thermometer registers 180 degrees in the middle of the meat loaf. Remove from oven and let rest while making the gravy.

Make the Tomato Gravy. Heat the bacon drippings in a 10- or 12-inch cast iron skillet. When the drippings are hot, add the onions and sauté until lightly browned. Add the flour all at once and stir to blend well; cook until slightly brown. Add the chicken broth and tomatoes, stirring to blend well. Season to taste with salt and pepper. Bring the mixture to a boil to thicken, then lower heat and simmer for about 5 minutes.

Slice meat loaf as desired and top each serving with some of the Tomato Gravy.

(Sandra Jullian)

BUTTERMILK PIE

The recipe for this sinfully delicious pie was given to Bonnie over thirty years ago by a "precious little old ranch woman." The crust is one you'll want to use often. Bonnie says you just can't have a good pie without a good crust.

Yield: Two 9-inch pies

Crust

1¾ cups all-purpose flour
1¼ sticks margarine at room temperature
1 egg, beaten
1 teaspoon vinegar
2 tablespoons ice-cold water

Filling

3½ cups sugar
2 tablespoons all-purpose flour
6 eggs, beaten
1 cup buttermilk
2 teaspoons vanilla extract
¼ pound (1 stick) margarine, melted

Make the crust. Blend the flour and margarine in a large bowl using a pastry blender until the mixture resembles coarse grain. Add the egg, vinegar, and cold water. Mix well, but don't overmix. Divide the dough into two equal portions and roll out on a floured work surface. Chill while preparing the filling. (Chilling prevents shrinking.)

Preheat oven to 300 degrees. To make the filling, combine the sugar and flour in a large bowl, tossing with a fork to blend. Add the beaten eggs, buttermilk, vanilla, and melted margarine. Stir mixture just until well blended; don't beat it. Pour into the chilled pie crust and bake in preheated oven for about 1½ hours, or until the filling doesn't jiggle in the middle when you shake it. Chill before slicing.

(Bonnie's Bakery)

Bonnie's Bakery

Leakey

Bonnie Crider, a dynamo of a lady, moved in 1970 from Houston to the Frio Canyon, where her husband's family owned Crider's Cabins, a fishing camp on the Frio River just south of Leakey. Bonnie opened her first restaurant, Country Kitchen, and became known for her delicious foods, especially her pies and other baked goods. ❄ In 1987 Bonnie took over management of Crider's Cabins and decided to give up the restaurant business, focusing on her first love in the kitchen: baking.

Thus began Bonnie's Bakery. In the early days she'd put a couple of pies in the oven, then run and clean a cabin. Over the years both businesses grew, and now she has three employees in the bakery, but she still manages the cabins. ❄ Folks start lining up outside the door of the tiny bakery before 7:30 in the morning. Bonnie will have the coffee on and plenty of hot doughnuts and her incredible cinnamon rolls, along with a display case full of other goodies. Stacked around the bakery you'll see parts of wedding cakes she'll deliver in the afternoon and various other special-order cakes, pies, cookies, and pastries. Bonnie will also ship her tempting treats (see the Web site for details).

❄ She said the best compliment she ever received was from a man who visited the bakery often. He said, "Bonnie, I don't know if you're making any money here, but you sure are making a lot of folks happy."

U.S. 83 (between RR 1120 and RR 1050) • (830) 232-5584 • www.cridersonthe-frio.com/bakery.htm

Frio Canyon Lodge Restaurant

Leakey

The Frio Canyon Lodge, located in downtown Leakey, has been a Hill Country destination for more than fifty years. The lodge was originally built in 1941 of native limestone and cedar. ❧ Connie and Walter Tidwell purchased the lodge and restaurant in 2006 after Walter retired from Continental Airlines as an airline pilot. Both were area natives, raised in Uvalde, about 40 miles south of Leakey. Connie had been in the catering and restaurant business for twenty-three years. The Frio Canyon Lodge Restaurant is the third and largest restaurant she's owned. It was by a stroke of fate, albeit not a great one, that the Tidwells came to buy the lodge. ❧ The last of their five children was in college when they decided to sell their home and transfer to New York City with Continental so they could travel. The day before the transfer papers came through, Walter had a stroke. Instead of traveling, they spent the next six months in limbo. Then they heard that the Frio Canyon Lodge was for sale and somehow knew that, indeed, "everything happens for a reason." They purchased the old lodge and completely refurbished it, creating a cozy, down-home feel. Connie developed a menu of taste-tempting Hill Country fare with a hefty Southwestern touch. The restaurant features a varying selection of wines, including some Texas labels, and Texas and Mexican beers. Enjoy dining on the lovely outdoor patio and breathe in that pure, clean, restorative Hill Country air. ❧ The Tidwells love living in Leakey, a booming little community of interesting and friendly folks. The area is a popular getaway destination, with the crystal-clear waters of the Frio River and Garner State Park nearby. The Leakey area has been listed in Texas Monthly's "Top Ten Hill Country Destinations" and is a favorite region with motorcycle riders, who come to ride the hills and hairpin curves of the "Three Sisters," Ranch Roads 335, 336, and 337 out of Leakey.

U.S. 83 at RR 337 • (830) 232-6810 • www.frio-canyonlodge.com

SOUTHWEST CHOPPED STEAK WITH LODGE QUESO AND JALAPEÑO BOTTLECAPS

Yield: 6 servings

6 (8-ounce) chopped sirloin steaks, seasoned with salt and black pepper to taste

Lodge Queso

¼ pound (1 stick) butter

2 Roma tomatoes, seeded and diced

3 serrano chilies, seeds and veins removed, minced

1 yellow onion, diced

3 large garlic cloves, minced

1 teaspoon ground cumin

1 pound American cheese, cut into small dice

2 cans (10-ounce) Ro-tel Diced Tomatoes & Green Chilies

1 can (6-ounce) evaporated milk

Jalapeño Bottlecaps

1½ cups well-drained sliced pickled jalapeños

2 eggs, beaten into 2 cups milk

All-purpose flour, seasoned with salt and black pepper

Canola oil for deep-frying, heated to 350 degrees

Make the Lodge Queso. Melt the butter in a heavy 3-quart saucepan over medium heat. Sauté the tomatoes, serrano chilies, onion, garlic, and cumin until the onion is wilted and transparent, about 6 minutes. Add the cheese, canned tomatoes with green chilies, and evaporated milk. Cook, stirring often, until the cheese has melted, about 15 minutes. Do not allow the mixture to boil. Set aside and keep warm.

Make the Jalapeño Bottlecaps. Dredge the sliced jalapeños in the egg wash, then dust with the seasoned flour and deep-fry in batches, taking care not to crowd the oil, or the chilies will stick together. Drain on paper towels on a wire rack set on a baking sheet; place in a warm oven while grilling the steaks.

Grill the steaks to the desired degree of doneness. Place a steak on each serving plate and top with a portion of Lodge Queso. Pile some of the Jalapeño Bottlecaps on top and dig in!

(Frio Canyon Lodge Restaurant)

CHICKEN AND SAUSAGE GUMBO

You can vary the taste of this delicious gumbo by using duck meat as a substitute for the chicken.

Yield: 8 to 10 servings

3 cups Gumbo Roux (see recipe below)

2 medium onions, chopped

2 large green bell peppers, chopped

3 large celery stalks, chopped

2 tablespoons minced flat-leaf parsley

2 teaspoons minced fresh thyme

½ teaspoon freshly ground black pepper

3 fresh bay leaves, minced

1 teaspoon minced fresh marjoram

2 heaping tablespoons filé powder

1 pound CrawLinks sausage links

2 quarts homemade duck or chicken stock

1 cup dry vermouth

2 pounds cooked chicken meat, cut into bite-size pieces, or substitute duck meat

1 teaspoon red (cayenne) pepper, or to taste

Salt to taste

Cooked white rice

Sliced green onions for garnish

Gumbo Roux

3 cups solid vegetable shortening or lard

3 cups all-purpose flour

Preheat oven to 350 degrees. Make the Gumbo Roux. Heat shortening in a heavy, deep-sided 12-inch skillet over medium heat. When the fat is hot, add the flour all at once and whisk until well blended and smooth. Continue to cook, whisking constantly, until roux is a deep mahogany color, about 30 minutes.

Add the vegetables and all seasonings except cayenne and salt. Cook the vegetables, stirring often, until they are very wilted and the onions are translucent, about 15 minutes. Remove from heat and set aside. While the vegetables are cooking, arrange the sausage links on a baking sheet and bake in preheated oven for about 10 minutes, or until the fat has been rendered. Pat the sausage links with paper towels to remove all traces of fat, then slice into bite-size pieces and set aside.

Bring the stock to a full boil in a 10-quart soup pot. Add the roux and vegetable mixture to the boiling stock; boil vigorously for 10 minutes, stirring frequently. Stir in the vermouth, chicken, and cooked sausage and boil for an additional 10 minutes. Lower heat and season with cayenne and salt. Cover and simmer for 45 minutes.

To serve, spoon about ⅔ cup cooked white rice into each serving bowl; ladle the gumbo over the rice. Garnish with green onions. Serve hot.

(Terry Thompson-Anderson)

Texas Cajun Sausage Company

Leakey

John L. Price is a really likeable guy who owns a ranch outside of Leakey. He considers himself primarily a rancher, but back in the spring of 2003, he began playing around with recipes for sausage. Eventually he came up with a sausage made from crawfish tails and pork. Everybody loved it, so John formed the Texas Cajun Sausage Company to sell and distribute the sausage under the trade name CrawLinks. Then he found a USDA-inspected facility to produce it in commercial quantities according to his recipe. ✸ Chances are, if you attend many festivals in Texas, you'll run into John and his CrawLinks. You can also order fully cooked, frozen CrawLinks from the Web site. The site also provides a list of the festivals where you can taste and purchase the sausage. John plans to add more retail outlets for the sausage in the future. ✸ CrawLinks is a really bodacious sausage. It's good and spicy and has just the right ratio of meat to fat. It's perfect in gumbo, on a sausage po'boy, or just to grill and eat.

1520 Old Reagan Wells Road • (830) 232-4424 • www.gocajun.com

Berry Street Bakery

Llano

"Llano was kind of like the Bermuda Triangle for us. It just sucked us in and we couldn't get out," says Christine De Lapp, who owns Berry Street Bakery with her husband, Frank. ❦ After the De Lapps bought a ranch in Mason County, they decided they really wanted to dabble in some sort of enterprise. Frank has long been an accomplished, self-taught baker, so the couple settled on Llano for the bakery after finding a charming old house on Berry Street across from the courthouse. ❦ Berry Street Bakery opened its doors in 2004 with the assistance of long-time friend and associate Addie Pinckney and two other employees. The large and inviting front porch overlooking the town square became an instant gathering place. ❦ Christine says that the items offered at the bakery evolved with customer requests. "Our customers began to ask us if we could make this or that—things their grandmothers used to bake. We never say no!" Not even to baking a cake that was a life-size replica of John Wayne for a Llano festival! ❦ There's an inviting array of kolaches, pastries, muffins, cookies, and cakes. The cakes made for special occasions are works of art, as are the hand-cut and beautifully decorated cookies that have been shipped all over the country. ❦ The bakery's signature breakfast dish is the Walkaway Omelet, an amazing sort of popover! The popular lunch menu consists of sandwiches served on homemade breads, salads, Frank's homemade soups, and the special of the week. ❦ The customers regard Berry Street Bakery as a local

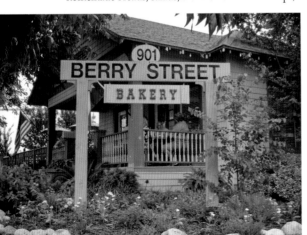

treasure. Some admit to eating there two or three times a week. It's a friendly place, with oilcloth tablecloths and old-fashioned salt and pepper shakers on the tables. The coffee pot's always on—and it's very good coffee. The De Lapps' motto, which they've posted on the wall, sums up the place: "Because Nice Matters."

901 Berry Street • (325) 247-1855

HOUSE GRANOLA

Berry Street Bakery sells this delicious granola in small bags displayed on top of the bakery counter. It's easy to make and would make great hostess gifts or Christmas gifts, sealed in zip-sealing bags and packed inside colorful tins.

Yield: About 1 gallon

1 cup dried cranberries
1 cup dark raisins
1 cup golden raisins
1 cup mixed dried fruit
¾ cup canola oil
1½ cups honey
1 teaspoon cinnamon
4 cups rolled oats
4 cups wheat flakes
2 cups sunflower seeds
2 cups raw wheat germ
1 cup chopped pecans
1 cup sliced unskinned almonds
1 cup whole almonds
1½ cups coconut

Preheat oven to 325 degrees. Combine the fruits in a bowl and toss to blend; set aside. Combine the canola oil, honey, and cinnamon in a bowl, whisking until well blended; set aside. In a large bowl combine all remaining ingredients. Pour the oil mixture over the dry ingredients in the large bowl. Stir to mix well. Spread the mixture out in a thin layer on baking sheets. Toast in preheated oven for 10 to 15 minutes, or until light golden brown. Do not overcook. Remove from oven and cool on wire racks. When completely cool, turn the mixture out into a large bowl and mix in the reserved fruits. Store in an airtight container.

(Berry Street Bakery)

ARUGULA AND BIBB LETTUCE SALAD WITH FIGS, GOAT CHEESE, AND HONEY-LEMON VINAIGRETTE

The vinaigrette for this salad can be made a day ahead and refrigerated.

Yield: 4 servings

Equal portions arugula and Bibb lettuce for
  4 portions, torn into bite-size pieces
12 fresh figs, halved (or quartered if figs are large)
1 cup crumbled semi-hard goat cheese
Pecan pieces

Honey-Lemon Vinaigrette

1 teaspoon minced lemon zest
¼ cup freshly squeezed lemon juice
¼ teaspoon kosher salt or fine-grain sea salt
2 tablespoons Fain's Honey
3 tablespoon chopped fresh chives
½ cup extra-virgin olive oil
Freshly ground black pepper to taste

Make the Honey-Lemon Vinaigrette. Combine all ingredients except the olive oil and pepper in work bowl of food processor fitted with steel blade. Process until well blended. With processor running, add the olive oil in a slow, steady stream through the feed tube. Add the black pepper and process for about 15 additional seconds. Transfer to a storage container and refrigerate until ready to serve.

To assemble the salad, toss the arugula and Bibb lettuce together. Place equal portions of the mixed greens on individual salad plates. Arrange some of the figs on each salad. Drizzle a portion of the Honey-Lemon Vinaigrette over the salads, scatter ¼ cup of the crumbled goat cheese over each, and garnish with a few pecan pieces.

(Terry Thompson-Anderson)

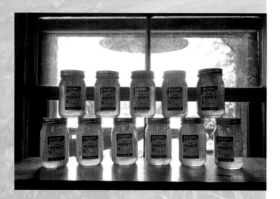

Fain's Honey

Llano

Fain's Honey is a small, unassuming place two miles south of Llano. Here you can purchase all of the company's products, but you'll also find them in supermarkets throughout the state. ❋ The company was established in 1926 by H. E. Fain, who kept several hives and began selling honey from unlabeled fruit jars. He was meticulous about the cleanliness and appearance of his honey and soon gained a reputation as a supplier of top-quality honey from 175 colonies of bees. During the Depression, when sugar was a luxury, he found a ready market for his honey. ❋ In 2005 Keith Fain, a grandson of the company's founder, acquired Fain's Honey, becoming the third generation of Fains to produce honey. Keith and his wife, Debra, expanded the product line to include raw native honey, honey butter spreads (such as a delicious jalapeño variety), ribbon cane

syrup, and sorghum molasses, all of which are available through the Web site. ❋ Today Keith has about 1,000 bee colonies, but the products are produced in a small-scale, hands-on operation. He doesn't feed his bees any substances to lengthen the production season. The company's motto is "Fain's honey doesn't pretend to be a gourmet honey, rather just a wholesome, natural Texas honey product." There are two key words there that sum up why it's so good—natural and Texas!

3744 State Hwy. 16 South • (325) 247-4867 • www.fainshoney.com

Llaneaux Seafood House

Llano

Robby Puryear, co-owner of Llaneaux Seafood with his wife, April, is an avid fisherman as well as a devotee of the food and culture of South Louisiana, hence the play on the spelling of Llano in the name of his restaurant, which specializes in Cajun seafood dishes. ❋ Robby and April grew up in Houston and have extensive backgrounds in the food business. They wanted to own their own restaurant, and in 2000 they

started to look for an ideal location, both for a restaurant and a home. They settled on Llano as a central location in the Hill Country with good growth potential and still affordable real estate. ❋ They found the perfect place for the Llaneaux Seafood House in a lovely 1903 Victorian home, located in a picturesque setting on the banks of the Llano River. They added a commercial kitchen onto the back of the house and remodeled the interior, turning the house into a charming restaurant reminiscent of South Louisiana's Cajun country. The wraparound porch overlooks the river, just beckoning one to sit a spell. It's a popular dining spot that fills up quickly in the evenings. ❋ The menu is one that's pretty unique in the Hill Country, featuring Cajun specialties like Spicy Crawfish Étouffée, Shrimp Étouffée, Seafood Gumbo, and Red Beans and Rice. Grilled fresh fish dishes are also available, such as mahi-mahi, salmon, and tuna, as well as fried seafood—shrimp, catfish, oysters, crawfish, frog legs, and alligator. Robby has created a Cajun version of the chicken-fried steak, topped with either cream gravy laced with andouille sausage or étouffée. We're talking the

real Tabasco here with this food! ❋ Regulars travel to the Llaneaux Seafood House from as far away as San Antonio for a good Cajun food fix in the heart of the Hill Country.

102 Legion Drive • (325) 247-3663 • www. llaneauxseafoodhouse. com

CRAWFISH ÉTOUFFÉE

Yield: 8 to 10 servings

1 pound (4 sticks) butter
4 cups chopped onion
2 cups chopped green bell pepper
2 cups chopped celery
¾ tablespoon minced garlic
1 teaspoon freshly ground black pepper
1 teaspoon red (cayenne) pepper
1 teaspoon crushed red pepper flakes
1 teaspoon paprika
½ cup Dark Roux (see recipe below)
1 cup all-purpose flour
1½ quarts shrimp or seafood stock, or substitute bottled clam juice
Kosher salt or sea salt to taste
4 pounds cooked, peeled crawfish tails, or substitute peeled, deveined raw medium shrimp
Cooked white rice

Dark Roux

½ cup canola oil
½ cup all-purpose flour

Make the Dark Roux. Heat canola oil in a heavy 10-inch skillet over medium-high heat. Whisk in the flour all at once and continue to cook, whisking constantly, about 25 to 30 minutes, or until roux is a dark, chocolate color. (If small black specks appear, the roux has burned and you must start over.)

Melt the butter in a heavy Dutch oven over medium heat. Add the onion, bell pepper, celery, and garlic. Cook, stirring often, until onions are limp and transparent, about 15 minutes. Stir in the seasonings, blending well. Add the Dark Roux and cook until it is dissolved and blended. Add the flour and blend well. Add the seafood stock and salt to taste; whisk until thickened. Simmer on medium-low heat for 15 minutes. Add the crawfish tails and cook only until they are heated through; avoid overcooking. (If using shrimp, cook just until they turn a rich coral-pink color.) Serve over white rice.

(Llaneaux Seafood House)

COCONUT MERINGUE PIE

Yield: One 9-inch pie

One baked 9-inch pie crust

Filling

1 cup sugar
¼ cup cornstarch
¼ teaspoon salt
3 cups whole milk
4 egg yolks (reserve whites)
3 tablespoons butter
1½ teaspoons vanilla extract
1 (3½-ounce) can sweetened flaked coconut

Meringue

4 egg whites at room temperature
¼ teaspoon cream of tartar
½ cup sugar
½ teaspoon vanilla extract
Toasted coconut

To make the filling, combine the sugar, cornstarch, salt, and milk in a heavy 3-quart saucepan over medium heat. Cook, stirring constantly until the mixture boils and thickens, then cook for an additional 2 minutes. Remove from heat and set aside. In a separate bowl, beat the egg yolks. Whisking vigorously, slowly stir 1 cup of the hot milk mixture into the yolks. Stir the warmed yolks into the remainder of the milk mixture and bring to a gentle boil. Cook, stirring, for 2 minutes. Remove from heat and add the butter and vanilla, stirring until the butter has melted. Add the coconut, blending well. Pour into the baked pie crust and set aside.

Preheat oven to 350 degrees. Make the meringue. Using an electric mixer, beat the egg whites and cream of tartar at high speed until foamy. Gradually add the sugar, 1 tablespoon at a time, until stiff peaks form and the sugar is dissolved, about 2 to 4 minutes. Add vanilla and beat just to blend. Spread the meringue over the hot filling, sealing it to the edge of the pastry. Sprinkle with toasted coconut. Bake in preheated oven for 12 to 15 minutes, or until lightly browned. Cool completely before slicing.

(Blue Bonnet Café)

Blue Bonnet Café

Marble Falls

The Blue Bonnet Café is an institution, having been a fact of daily life in Marble Falls since 1929. It's one of those places where the locals gather and everybody seems to know the person at the next table. ❧ Originally located on Main Street, the café moved to its present site "on the highway" in 1946. It quickly became a favorite stopping place for folks who were just passing through town. Today you just might see a famous country western singer, movie star, NFL great, or other celebrity bellied up to the counter or settled into a booth, chowing down on the famous grub for which the café has always been known. The café has been written up in hundreds of publications from coast to coast. ❧ John and Belinda Kemper bought the Blue Bonnet in 1981 and set about to not change a thing! They had been in the hospitality and food business for most of their lives and were thrilled to own such a legend. ❧ When the Kempers' daughter, Lindsay, was an exchange student in high school, she took some Blue Bonnet Café t-shirts to her host family in Australia. The father was working on a project in Antarctica and one day wore his t-shirt to work. Another man pointed to the shirt and said, "The Blue Bonnet? I'm from Spicewood, Texas, and I eat there all the time when I'm home." ❧ The staff is very much like a big, happy extended family. The waitresses are some of the friendliest you'll find anywhere—despite the fact that on an average day the Blue Bonnet serves 2,000 meals! And they're open seven days a week. Breakfast is served all day, and a daily hot lunch special is served from 11 a.m. until closing time. Dinner is served every day except Sunday. It's true comfort food, complete with plate-size chicken-fried steak with cream gravy and mashed potatoes. Big, puffy yeast rolls are served with hot meals, and they'll keep bringing them as often as you ask! ❧ But the real draw at the Blue Bonnet is the impressive array of homemade pies of every description, displayed in the pie case at the entrance to the dining room. You can buy a slice or the whole pie. They're as good as they look, too. The Coconut Meringue is by far the most popular. ❧ Lindsay and her husband, David, are taking over the reins at the café, giving Mom and Dad a little breather. But you'll still find John there most mornings, shaking hands with regulars and greeting the first-timers.

**211 U.S. 281 • (830) 693-2444 •
www.highlandlakes.com/bluebonnet**

Brothers Bakery & Café

Marble Falls

Brothers Bakery & Café, which opened in 2002, was the culmination of a lifetime dream. Chef Ryan LeCompte Malamud, the owner, was born in New Orleans to a family whose life revolved around great food. The family moved to Georgetown, Texas, when Ryan was six. As a high school senior, Ryan was one of the first students to enroll in a newly offered culinary class, fueling his growing interest in pursuing a career in food. After graduation he traveled extensively in Europe, working in bakeries in Italy, France, Spain, and the Netherlands, learning the techniques of fine European baking as well as bakery operations. After returning to the United States he earned a degree from the Culinary Institute of America in Hyde Park, New York, and then went on to pursue a degree in baking and pastry arts from the Institute's Napa Valley campus in 2001. ❈ Ryan named the bakery in honor of his brother Kurt, who lives in Louisiana but works with Ryan at the bakery several times a year when he comes to visit. ❈ The bakery makes pastries, muffins, breads, cookies, pies, quiches, and the most incredible French-style croissants you'll find in the Hill Country. Ryan worked painstakingly on the recipe to achieve a perfect European-style pastry. He uses real butter in all of the pastries, never manufactured, "fake" fats. The cakes are not only luscious but beautiful creations made with real buttercream icing. The bakery serves lunch each day—soups, sandwiches, and innovative salads, all made from scratch. The sandwiches, naturally, are served on housemade artisan breads. It's a busy place with lots of hustle and bustle in the open kitchen behind the pastry case. There are racks of fresh breads of every description. The pastry case itself is about 8 feet of temptation, crammed with mouth-watering goodies. ❈ You'll run into Ryan at restaurants and resorts all over the Hill Country as he makes deliveries for his thriving wholesale business, which provides baked goods to fine dining establishments.

519 U.S. 281 North (at 6th Street) •
(830) 798-8278 • www.brothersbakery.com

AMY'S FAVORITE COOKIES

Ryan named these really great cookies after his cousin Amy, who thinks they're great, too.

- 1 cup butter
- 2 cups sugar
- 3 eggs
- ½ teaspoon allspice
- 1 teaspoon ground cinnamon
- 1 teaspoon baking soda
- 1 teaspoon vanilla extract
- 2 teaspoons salt
- 1 teaspoon baking powder
- 2 cups chopped pecans
- 5 cups all-purpose flour

Preheat oven to 350 degrees. In bowl of electric mixer, cream the butter and sugar together until fluffy and light in color, 4 to 5 minutes. Add the eggs, one at a time, scraping down the sides of the bowl after each addition. Add spices, baking soda, vanilla, and salt, beating just to blend. In a separate bowl combine the baking powder, pecans, and flour, whisking to blend well. Add the flour mixture to the butter mixture and blend well, scraping down the sides of the bowl often. Drop the batter by teaspoonfuls onto a cookie sheet. Bake in preheated oven for 10 to 12 minutes. Cool on a wire rack.

(Ryan LeCompte Malamud)

Flat Creek Estate
Marble Falls

Rick and Madelyn Naber jokingly say that they purchased this eighty-acre property, formerly a Limousin cattle ranch, because of its attributes, not because the location was convenient to anywhere! ❧ Wanting to maintain the land's agricultural character, the Nabers made exhaustive soil and water tests, all of which confirmed that the site—with its deep soils, natural springs, and sloping hillsides—was ideal for viticulture. Vines were first planted on six acres in 2000, and this has now expanded to twenty acres of vineyards. Along with such signature varietals as sangiovese, shiraz, and Portuguese port varietals, Flat Creek Estate produces noteworthy wines from muscat, tempranillo, and pinot grigio grapes. ❧ Flat Creek's winemaker, Tish Cooper, hails from the American West. She began studying fermentation science at Oregon State University–Corvallis, intending to pursue a career as a brewmaster. After classes in viticulture, however, she became enamored with the more hands-on aspects of viticulture and enology and abruptly changed directions. She studied winemaking in France, then obtained a degree in enology at California State University at Fresno. Before moving to Texas to make Big Country wine, Tish worked at a Southern California winery specializing in port and sherry wines. ❧ The Nabers constructed a 5,000-case winery in 2001, with a commitment to producing quality Texas wines under two labels. The Flat Creek Estate label is devoted to estate-grown fruit and includes Pinot Grigio, Muscato Blanco, Shiraz, and "Super Texan" Sangiovese wines. The Travis Peak Select label presents a broader range of wines, crafted entirely from Texas fruit, including Cabernet Sauvignon and Orange Muscat. Flat Creek wines have won many awards and accolades. A grand facility with an event center and tasting room, completed in 2005, offers a spectacular view of the vineyards and a perfect venue for wine releases, cooking classes, weddings, banquets, and other events.

24912 Singleton Bend East Road (off FM 1431) • (512) 267-6310 • www.flatcreekestate.com

ZUCCHINI AU GRATIN

This gratin may be served warm or cold. It's a great side dish with grilled or roasted white meats, especially when paired with a dry Texas rosé wine.

Yield: 6 servings

2 tablespoons olive oil

1 large onion, cut into ½-inch dice

2 garlic cloves, minced

1 fresh bay leaf, minced

1 teaspoon minced fresh thyme

9 medium zucchini, cut into ½-inch dice

Salt and freshly ground black pepper to taste

2 eggs, beaten

1 teaspoon cornstarch blended with 1 tablespoon cold water

⅛ teaspoon freshly grated nutmeg

6 ounces (1½ cups) shredded Swiss cheese

Preheat oven to 425 degrees. Butter bottom and sides of a 3-quart oval au gratin dish; set aside.

Heat oil in a deep, heavy-bottomed, 12-inch skillet over medium heat. When oil is hot, add the onion and garlic. Sauté until onion is wilted and transparent, about 5 minutes. Add the bay leaf, thyme, and zucchini. Cook, stirring frequently, until zucchini is very soft. Season with salt and pepper and remove from heat. Mash the zucchini mixture, blend in the eggs, and stir in the cornstarch mixture and nutmeg, combining well. Turn zucchini out into prepared au gratin dish and scatter the shredded cheese over the top.

Bake in preheated oven for about 15 to 20 minutes, or until cheese is browned and zucchini is bubbly.

Variation: You can use your creativity to vary this recipe. Substitute other herbs, trying basil, mint, herbs de Provence, parsley, etc. Also try using Parmesan cheese in lieu of Swiss, or use equal parts of both cheeses.

(Fleur de Ble French Bakery)

Fleur de Ble French Bakery

Marble Falls

When Jean-Claude and Béatrice Walter opened Fleur de Ble ("wheat flower") French Bakery in August of 2008, they weren't using the term "French bakery" loosely. The couple immigrated to the United States from Lyon, France. ❧ Jean-Claude had a twenty-four-year career in the French gendarmerie with stints in Corsica and then Guadeloupe, where Béatrice had the opportunity to train as a baker. After returning to France, the pair made a trip to the Texas Hill Country to see its legendary bluebonnets. Impressed by the town of Marble Falls, they made the decision to move here and open the kind of bakery found everywhere in France. Fleur de Ble has enjoyed a warm reception from local residents who enjoy the tempting array of fresh French pastries, beautifully glazed fresh fruit tarts, artisan breads, and much more.

❧ Fleur de Ble is a charming spot for breakfast or lunch. In the French tradition, the coffee is fresh, hot, and deliciously bold-flavored, served, of course, with real cream. The dining area is cozy, with a distinctly European ambiance. Lunch specials change daily, offering delicious pairings of classic French dishes. There's always a selection of individual quiches with various fillings and divine pastry, and excellent sandwiches prepared on small French breads. ❧ When questioned about the challenges of baking the French way in Texas, Béatrice explained that American flour is much higher in gluten than the flours milled in France, making tender pastries more difficult to achieve, especially with American butter and its lower butterfat content. However, she's done a fabulous job of compensating, as the croissants are golden brown works of art, with hundreds of layers of flaky pastry and a wonderful buttery flavor. Her quiches and tarts are light and perfectly tender. ❧ A small section of the bakery features imported French candies and sweets. ❧ The Walters enjoy explaining the delicacies of French bakery fare to their eager customers. They, along with their son, are enjoying the region's wide open spaces and welcome hospitality. And each spring there are those bluebonnets that brought them here.

809 Twelfth Street • (830) 265-4738

BLACKBERRY CROSTADA WITH STREUSEL TOPPING

A crostada is a free-form pastry that can be filled with many kinds of fruit and berries. They are really quick and easy to make. Don't worry about the final shape of the crostada—the rustic look is part of its homemade charm. It's an all-around winner when made with fresh-from-the-vine Texas blackberries.

Yield: 6 to 8 servings

Chilled whipping cream

Pastry

1 cup all-purpose flour
Pinch of salt
¼ pound (1 stick) frozen unsalted butter, cut into 1-inch cubes
3 to 4 tablespoons very cold water

Filling

1 pint (2 heaping cups) fresh blackberries
1 teaspoon finely minced lemon zest
⅔ cup sugar
½ cup raspberry liqueur

Streusel Topping

3 tablespoons unsalted butter, cut into small cubes
1 teaspoon vanilla
½ cup all-purpose flour
½ cup sugar

To make the pastry, combine the flour, salt, and butter in work bowl of food processor fitted with steel blade. Pulse until the butter is broken up into pea-size bits. Add 3 tablespoons of the cold water and process until well blended. Check the consistency of the dough by gathering a small amount into your hand and squeezing. It should form a ball.

(continued)

Sweet Berry Farm

Marble Falls

Sweet Berry Farm is a family destination on Saturdays and Sundays. It's exactly what Dan and Gretchen Copeland envisioned when they started the twenty-acre "pick-your-own" farm outside Marble Falls in 1999. ❦ Here you can pick as many strawberries, blackberries, tomatoes, potatoes, pumpkins, and gourds in their respective seasons as you can haul away. The Copelands will even provide clippers and gloves for those thorny blackberry vines. Except for the pumpkins and gourds (which are priced individually), you pay for fruit by the pound. ❦ But be sure to pack a lunch. After Mom, Dad, and the kids have picked for a while, everybody can ramble down to the covered arbor and have a bona fide picnic. Don't worry about bringing dessert, though—the farm sells homemade ice cream, made with the farm's own fruit, as well as awesome smoothies and shakes. The homemade strawberry preserves are also mighty fine. ❦ You'll probably be met by Dan's father, Max Copeland, who is the official greeter at Sweet Berry Farm. He retired after forty years of being a pastor in Marble Falls. Max will certainly steer you to the Texas Maze, an immensely popular feature at the farm. It's a four-acre maze formed by 8-foot-high sorghum-sudan grass cut in the shape of the state of Texas. Within the maze are numerous paths and almost a dozen Texas "cities" at points that approximate their real geographical locations in Texas. There's a new game theme each spring and summer. You get cards revealing the theme and you have to find the cities—and then find your way out! Also for the kids there are goats, donkeys, and horses for petting.

1801 FM 1980 • (830) 798-1462 • www.sweet-berryfarms.com

If the dough is too dry to form a ball, add the remaining 1 tablespoon water, or more if needed. Process just until water is blended. (Don't allow the dough to form a ball in the processor.) Turn crumbly dough out onto a work surface and knead together three or four times to form a cohesive dough. There should still be chunks of butter in the dough. Pat dough into a disk, wrap in plastic wrap, and refrigerate for 20 minutes.

Preheat oven to 375 degrees. Line a 14-inch-wide baking sheet with parchment paper; set aside.

To make the filling, combine all ingredients in a medium bowl and stir to blend, taking care not to break up the berries. Set aside.

Make the Streusel Topping. Combine all ingredients in work bowl of food processor fitted with steel blade and pulse until the mixture is crumbly; set aside.

To assemble the crostada, turn the pastry out onto a floured work surface and roll into a 12-inch round. Roll the pastry around the rolling pin and unroll it onto the parchment-lined baking sheet. Using a slotted spoon, spoon the berry filling into the center of the pastry. Discard the juice from the bowl. Working quickly, fold the edge of the dough up around the berries in a pleated fashion, leaving about 4 to 5 inches of exposed berries in the center. Scatter the Streusel Topping over the entire crostada; bake in preheated oven for 30 to 35 minutes, or until the pastry is golden brown and the filling is bubbly and has formed a nice syrup. Slide the crostada onto a wire rack by lifting both sides of the parchment paper. Cool until lukewarm, then slice into wedges. Serve in shallow bowls, drizzling whipping cream over each serving.

(Terry Thompson-Anderson)

Zoo La La

Marble Falls

It was a case of perfect timing when Donald and Debby Columbus moved to Marble Falls in 1996. Debby had always loved the Highland Lakes region, having vacationed there in her childhood. Donald, a native of Canada, first visited Marble Falls on the couple's first wedding anniversary and loved the area at once. As their children grew up and moved away, Debbie and Donald realized that Marble Falls was a central location for their five children and grandchildren, so the move made sense. ❉ Debby had always loved to cook and, like most serious cooks, loved cookware and every sort of cooking utensil. Very shortly after moving to Marble Falls, she heard that the town's popular cookware store/cooking school was for sale and knew that she was meant to own it! ❉ The Columbuses streamlined the store's inventory, making sure every item has a purpose, whether to make cooking and entertaining easier and faster or to grace the table with the best in serving pieces, tableware, flatware, and glassware. Debby tests every kitchen implement before it goes on the shelf—no frivolous, useless gadgets here. The store also has a marvelously tempting selection of chocolates and other handmade candies. ❉ The attractive open classroom kitchen was a natural adjunct to the store, with popular classes taught by area chefs and cookbook authors. Debby and Donald also added a wine shop that specializes in hard-to-find wines from boutique wineries and provides a great selection of glassware and

accessories for the wine enthusiast. Donald, a certified sommelier, offers wine tastings on Saturdays and chooses a perfectly matched wine for each course in the cooking classes. ❉ Wine and specialty beers are available by the glass, a feature popular with husbands who get wrangled into joining their wives on shopping expeditions! Plans are in the works to add artisan cheeses and specialty meats to the store's slate of specialty foods. ❉ The Columbuses take pride in the strong customer base they've established through personal service. If you're looking for a special kitchen gadget or unusual serving piece, Debby loves to seek out these obscure, interesting items for customers. "It's what separates us from the big box stores," she says.

309 Main Street • (830) 798-0161 •
www.zoolala.com

Akashic Vineyard

Mason

"**S**nip at the top of the grape cluster—and no MOG" (material other than grapes), intoned Akashic Vineyard owner Don Pullum as the group of volunteer grape pickers headed to the 1.5-acre section of the vineyard planted in mourvèdre grapes, buckets and clippers in hand. ❋ Each year at harvest time volunteers show up from all over the Hill Country and from all walks of life to help with the harvest. They go down the rows singing songs and telling tales while filling five-gallon buckets with grapes. When all the fruit has been picked, it will be hauled to Sandstone Cellars Winery in Mason, where Don is also the winemaker. After being crushed, the grapes will go into the fermentation tanks where the process of making another vintage of fine wine will begin. ❋ How did an Ivy League–educated venture capitalist, with no family ties to agriculture, come to be the steward of seven and a half acres of grapevines? Don says that he began his lifelong love of wine at age sixteen when he was given a small pour of a Chablis Premier Cru from the estate of Albert Pic. He knew the wine was good but lacked the sensory recall and vocabulary to describe its bouquet. It simply tasted good. ❋ While working in Houston, Don met the late Camille Berman, the wickedly charismatic owner of Maxim's Restaurant. Camille liked Don and became his friend and wine mentor. ❋ In 1998 Don found the land for his vineyard and planted the first vines, an acre of primitivo grapes, followed in subsequent years by plantings of grenache, sangiovese, mourvèdre, and syrah. ❋ The first commercial wine that Don crafted was the 2004 Sandstone Cellars Syrah, released in November of 2005. It was a very good vintage. As of this writing, the vineyard is nine years old, and in November of 2007 the third vintage crafted from Akashic's grapes was released. ❋ The vineyard is not open to the public, but serious oenophiles may call to arrange a tour or to volunteer for picking.

Bluff Creek Road • (325) 265-4221

Sandstone Cellars Winery

Mason

Sandstone Cellars is a small boutique winery in one of the Hill Country's newest wine regions, Mason County, which is proving to be excellent for growing grapes. The Hickory Sands soil and moderate weather seem to be a particularly good match for growing Mediterranean grape varietals. ❧ Sandstone Cellars, the first winery to open in Mason County, was established by partners Manny Silerio and Scott Haupert in 2003 after state laws were passed allowing wineries to operate tasting rooms in dry counties. The winery is located in a charming raised cottage across from the courthouse on Mason's town square. ❧ Scott and Manny met while both were in college in San Antonio—Manny a business major and Scott a journalism major. Scott's real love, however, was the viola. Classically trained, Scott went on to receive a master's degree in music performance from Yale. The pair moved to Los Angeles, where Manny became a merchandiser for Gap clothing stores while Scott pursued a professional music career, playing on more than 300 film scores, including *Fried Green Tomatoes, Jurassic Park,* and *All the Pretty Horses.* Scott appeared twice on *The Tonight Show* with Jay Leno and has performed on over 100 recordings, including orchestral backup for Barbra Streisand. ❧ The winemaker at Sandstone Cellars is Don Pullum, who was the first to plant wine grapes in Mason County, starting with one acre of grapes on his Akashic Vineyard property in 1998. ❧ Sandstone Cellars released its first wine, a Syrah, in 2005, followed in 2006 by Sandstone II, a blend of eight Mediterranean varietals. Both wines have met with critical acclaim. Sandstone II was awarded a bronze medal at the 2007 San Francisco Wine Fair Competition, one of the country's most prestigious wine competitions, beating out some very respected California vintners for the title. Sandstone III, released in 2007, was another great vintage. ❧ With a track record of three out of three wines being far superior to the norm, it's a safe bet that we'll hear more great things about these guys. ❧ The winery is also a good place to purchase noteworthy wines from other Texas wineries, some of which do not have tasting rooms.

211 San Antonio Street (U.S. 87) • (325) 347-9463 • www.sandstonecellarswinery.com

Santos Taqueria y Cantina

Mason

During the lunch or dinner hour from Thursday through Saturday and at lunchtime on Sundays, there's standing room only at Santos Taqueria. This small taqueria and cantina—owned by Sandstone Cellars Winery partners Manny Silerio and Scott Haupert, along with Manny's mom, Santos—is a favorite with locals. Santos Taqueria y Cantina has been named as one of the Hill Country's 25 Best Places by *Texas Monthly* and praised in many other publications. ✤ Located on the town square in a refurbished gas station, the

taqueria is a unique place. There are a few tables and a couple of booths inside, plus a spacious covered patio outside where the gas pumps used to be. The grounds have been lushly landscaped, so it's enjoyable to sit outside and watch the goings-on in Mason. ✤ Santos Silerio not only runs the kitchen but also prepares most of the food herself. And it's all made from scratch, just the way her grandmother taught her. ✤ You'll find such sinfully delicious gems as gorditas with your choice of fillings. Gorditas are a deep-fried pocket made from corn masa and stuffed with various ingredients. They're hard to find at Mexican restaurants because they're so time-consuming to make—and they're only as good as the filling! ✤ The menu also features chalupas, taquitos, quesadillas, nachos, and a fabulous Chile con Queso. Whatever your preference, it's all great. ✤ Santos Taqueria makes four different salsas fresh each day, all distinctively and deliciously different from any other salsas. ✤ Scott and Manny often do double duty at both the winery and the taqueria, working behind the counter or cleaning tables. And Santos will be scurrying around in the open kitchen,

patting out gorditas or tending orders on the stove. ✤ You can purchase a bottle of wine at the winery and take it next door to the taqueria to enjoy with your lunch. Any one of the three Sandstone vintages would be perfect with an item from the taqueria menu. ✤ It's a busy little place, and that's part of its charm, coupled with the fact that Scott, Manny, and Santos are just good folks.

205 San Antonio Street (U.S. 87) • (325) 347-6140

SQUARE PLATE BREAD PUDDING WITH CUSTARD SAUCE

Yield: 10 to 12 servings

5 eggs

4 cups milk

2 teaspoons ground cinnamon

1 cup raisins

1 pound day-old baguettes, torn into bite-size pieces

Brown sugar

Custard Sauce

¾ cup sugar

3 egg yolks, lightly beaten

⅛ teaspoon salt

¼ cup water

3 tablespoons butter, cut into small bits

2 tablespoons cornstarch

1½ cups boiling water

2 teaspoons clear vanilla extract

Preheat oven to 350 degrees. Grease a 9-by-13-inch glass baking dish and set aside. Combine the 5 eggs, milk, cinnamon, and raisins in a large bowl; whisk to blend well. Add the baguette pieces and stir to combine. Turn out mixture into prepared baking dish and bake in preheated oven for 45 minutes, or until top of pudding springs back when lightly pressed. Cool on rack.

While the pudding is baking, make the Custard Sauce. Cream the sugar and egg yolks in mixing bowl until mixture is fluffy and a pale lemon yellow in color. Add salt, ¼ cup water, and butter. Beat until well blended. Add the cornstarch and beat to blend. Slowly add the boiling water, while beating, until well combined. Turn mixture out into a heavy-bottomed saucepan and cook over medium heat until thickened, stirring constantly. Add vanilla extract and blend.

Cut pudding into squares and serve warm, topped with warm Custard Sauce.

(The Square Plate)

The Square Plate

Mason

For almost thirty years, Larry and Barbara Brown were the successful proprietors of Engel's Deli, a popular breakfast and lunch spot in Fredericksburg. After the last of their seven children graduated from high school, Larry and Barbara were ready for a rest. They sold the deli and spent a year in New Mexico. However, the couple realized that Texas was really home, so they headed back. ❖ Having been in the restaurant business for so long, they weren't satisfied with being idle. So they decided to establish a small eatery in Mason by remodeling a vintage building they owned on the town square. With her flair for creating spaces with ambiance, Barbara transformed the building into a cozy, small-town café with the original storefront windows and glass doors and with a nostalgic feel that brings back memories of an earlier time. The aromas of good coffee, desserts fresh from the oven, and simmering soups beckon patrons to have a seat and open up the newspaper. ❖ The Square Plate, open for lunch only, has a tempting array of sandwiches, such as Grilled Reuben on Rye, good old Egg Salad, and a toasted goat cheese, pesto, and tomato sandwich served on ciabatta bread. The full-meal salads include Fruited Chicken Salad, the most popular item at the Browns' deli in Fredericksburg. The menu also offers homemade soups that change daily as well as a schnitzel plate and a German sausage plate, served with either kraut or red cabbage. The sides are all lovingly prepared, and the potato salad is a delicious, unique concoction with a vinegar and oil dressing. A daily Hot Lunch Special consists of classics like meatloaf or other comfort foods that won't disappoint. An old-fashioned glass deli case is filled to the brim with such classics as carrot cake, bread pudding, cheesecake, brownies, chocolate pecan bourbon pie, caramel apple pie, and other goodies. ❖ With Larry in the kitchen and Barbara behind the counter, the Browns love what they're doing, and the evidence comes on square plates. You won't leave hungry.

212 Fort McKavitt • (325) 347-1911 • www.sqplate.com

Tallent Vineyard

Mason

Drew Tallent is a fourth-generation Mason County farmer who knows the responsibilities that come with stewardship of the family farm. He's a man with a remarkably serene temperament, very much in tune with nature and all its nuances. He's raised watermelons, peanuts, wheat, sesame, guar beans, cantaloupes—whatever crops would keep the farm profitable and in the family. ❄ When wine grapes began to make the Texas scene on a large scale, Drew researched the profitability of growing grapes and decided this crop might work for the farm. He planted four acres of cabernet sauvignon, and the vines flourished. The following year, he planted eighteen additional acres in syrah, malbec, and grenache. Drew had met Richard Becker, owner of Becker Vineyards, who had indicated a need for malbec in Texas. So Talent Vineyard grew the grapes for Becker Vineyard's first Malbec, released in late spring of 2007 to great acclaim. Eventually Drew also planted

pinot grigio, merlot, mourvèdre, and petit verdot varietals. ❄ He learned to fight the late-spring freezes so common in the area by installing one of the first sprinkler systems in the Hill Country to completely cover the vines with a layer of ice during the freeze. When the freeze is over, the ice melts

and the tiny buds keep right on growing. In 2006, however, the grapes couldn't survive the three vicious hailstorms that pounded the fields, also wiping out that year's peach and berry crops. ❄ With his dry sense of humor, Drew jokingly says, "You know how to make a small fortune in the grape-growing business? Start with a large one!" ❄ On any given day, you'll see Drew working somewhere on the farm. There's nothing easy about keeping tractors, sprayers, pumps, and drip systems working or making sure that all the pruning, hedging, and tucking is done on the vines and all the rows are mowed. It's a continuous process to make the finest grapes possible on the Hickory Sands of Mason County. ❄ The vineyard is not open to the public, but serious oenophiles may call to arrange a tour.

(325) 258-4489

Love Creek Orchards

Medina

Baxter and Carol Adams, owners of Love Creek Orchards, are special folks who have given Texas a couple of real treasures. In 1981 Baxter and Carol purchased 1,863 acres along Love Creek in the Bandera Canyonlands west of Medina. The land was totally depleted, but they fell in love with its rugged beauty. ❈ Paying no heed to those who told him that apples wouldn't grow in the Texas Hill Country, Baxter began experimenting with planting

dwarf apple trees that produce full-size fruit and ripen early. His success gave rise to a flourishing apple-growing industry around Medina. ❈ In the mid-1990s the Adamses planted a second apple orchard, located in Medina. Because they are tree-ripened, Love Creek's apples contain as much as 40 percent more sugar than commercially grown apples from other states. Love Creek Orchards grows fourteen varieties of apples. The Gala apples beat others to market by a good three weeks. ❈ Love Creek Orchards also grows thornless Ouachita blackberries. Both the apples and blackberries are available as pick-your-own during the season. ❈ The Cider Mill and Country Store offers various varieties of apples picked fresh from the trees, fresh apple cider, and just about anything that can be made from apples. You can even buy the same kinds of apple trees and blackberry vines that are grown in the orchards as well as bigtooth maple trees, which Baxter helped bring back from near extinction in the Hill Country. ❈ The kitchen at the Cider Mill turns out washtub-size

apple pies made from Pink Lady apples as well as apple butter, jams, jellies, and homemade breads. Many of these products, as well as fresh-picked apples, can be ordered from the Web site. ❈ The Patio Café at the Cider Mill offers daily lunch specials, plus a regular menu of tempting dishes and sandwiches—and, of course, slices of those gargantuan pies. The café will even pack a picnic lunch that you can enjoy under the great oaks beside the pond at the orchards. ❈ In 2000 the Adamses created a second treasure for Texans to enjoy by selling 1,400 acres of their Love Creek Ranch to the Nature Conservancy to create the Love Creek Preserve.

14024 State Hwy. 16 North • (830) 589-2588 • www.lovecreekorchards.com

APPLE BUTTERMILK LOAVES

Yield: 2 loaves

3 cups all-purpose flour
½ teaspoon baking powder
½ teaspoon baking soda
¼ teaspoon salt
½ pound (2 sticks) butter
1 cup granulated sugar
1 cup firmly packed brown sugar
3 eggs
1 teaspoon vanilla
¾ cup buttermilk
½ pound apples, peeled, cored, and finely chopped (1½ cups)
½ cup chopped pecans
1 teaspoon grated orange zest

Preheat oven to 350 degrees. Grease and flour two 8½-by-4½-by-2⅝-inch loaf pans; set aside. Combine the flour, baking powder, baking soda, and salt in a large bowl, tossing to blend well; set aside. In bowl of electric mixer, cream the butter, sugar, and brown sugar; beat in the eggs one at a time, scraping down the sides of the bowl after each addition. Add vanilla; beat just to blend. Add the buttermilk alternately with the flour mixture, scraping down the sides of the bowl after each addition. Add the apples, pecans, and orange zest. Beat just to blend. Divide the batter between the two prepared loaf pans. Bake for 50 to 60 minutes, or until a toothpick inserted near the center comes out clean. Cool in the pans for 10 minutes, then turn out loaves onto a wire rack and cool completely. Wrap and let stand overnight to allow flavors to blend.

(Love Creek Orchards)

STEWED SQUASH
WITH SHRIMP AND SPINACH SOUP

Carol Taylor says that her partner Chef Mike Hennessey is perhaps best known for his Soup of the Day. "Who would have thought that West Texans would eat soup all year round?" she asked. "We certainly underestimated the simple practical comfort of a meal of homemade soup!" Carol says many people have driven quite a distance to get some of this soup.

Yield: 4 to 6 servings

¼ pound (1 stick) butter

1 large yellow onion, chopped

4 celery sticks, chopped

2½ pounds yellow squash, chopped

2 cups chicken broth

3 ounces fresh baby spinach, chopped fine

½ pound small (70–90 count) shrimp, peeled and deveined

½ teaspoon freshly ground white pepper

Salt to taste

1 cup whipping cream

Melt the butter in a heavy 3-quart saucepan over medium heat. Add the onion, celery, and squash. Sauté, stirring often, until the vegetables are wilted and very tender, about 20 minutes. Remove from heat and puree the vegetables in a blender until smooth. Return the puree to the pan and add all remaining ingredients. Stir to blend well and cook until the shrimp have turned a rich coral-pink color, about 15 minutes. Serve hot.

(Sideoats Café & Bakery)

Sideoats Café & Bakery

Menard

When two of Menard's restaurants closed in 2002, the small town really felt the loss—especially the devoted bridge players who had lost their game venues. Carol Taylor decided that the community needed a gathering place with good, fresh, affordable food and set about to open one. Carol also figured that since Menard is the eastern gateway to the Great Plains, the restaurant would be a good stopping place for hungry travelers. ❧ Two years later, in 2004, Carol opened the Sideoats Café & Bakery. She named it for sideoats grama, which was designated the state grass of Texas in 1971. A nutritious, drought-tolerant grass that is highly prized by ranchers like Carol's husband, the grass seemed to be a fitting name for a café emphasizing good, nutritious food. ❧ Carol first met Chef Mike Hennessey when he was working for a well-know catering firm in San Angelo. She'd loved his food, and in 2005 they became partners in Sideoats. Mike describes himself as a "street chef"—completely self-taught. He's done a marvelous job of developing the "upscale down-home" menu, which offers a diverse selection of items, including a very worthy chicken-fried steak. Mike says his Grilled Flat Iron Steak, topped with a blue cheese–scallion butter, is becoming one of the most popular items on the menu. ❧ Sideoats is open for breakfast, lunch, and dinner. Lunch offers a variety of cold and grilled sandwiches served on house-baked LaBrea breads, delicious burgers, and Mike's great soups. The bakery makes fabulous cinnamon rolls and other breakfast pastries, muffins, and unique desserts. Fresh-baked LaBrea breads can be purchased by the loaf.

509 Ellis Street (U.S. 83) • (325) 396-2069 • www. sideoatscafe.com

Huisache Grill and Wine Bar

New Braunfels

Huisache Grill opened in 1994 in a building that had a long history as a restaurant site in historic downtown New Braunfels. Owners Don and Lynn Forres jokingly say that it "had more holes in its walls than wall, little street presence, and the charm of an active railroad track fifteen feet from the building." ❋ But through a remodeling and expansion project in 1998, Don and Lynn created a casual, unique restaurant offering moderately priced but

innovative lunches and dinners with an extensive beer and wine list. ❋ Each of the three dining areas resulting from the renovation and expansion has its own distinctive personality. The small front dining room has a sleek, modern feel with soft coloring. The large dining room and bar area has a rustic feel with a high vaulted ceiling and exposed rustic beams. A rear dining room has a homey, country-café look. ❋ Huisache Grill boasts friendly, prompt, and caring service, plus great food. The menu has a wide and varied selection of salads, including the outstanding Asian Salmon Salad Plate, as well as soups, really innovative sandwiches, grilled steaks, seafood dishes, and other

seafood items. The menu also features pork chops, pasta, chicken dishes, and a very good version of the Hill Country mainstay, chicken-fried steak. Desserts are all made in the restaurant kitchen and include a superb Italian Cream Cake.

303 W. San Antonio Street • (830) 620-9001 •www.hui-sache.com

SHRIMP AND CRAWFISH BISQUE

Yield: 12 servings as soup course,
6 servings as entrée

½ cup butter
¼ cup minced onions
¼ cup minced celery
½ cup flour
½ gallon milk
2 tablespoons chicken base
½ teaspoon ground white pepper
½ pound cooked cocktail shrimp
1 pound crawfish tails
2 cups dry sherry (not cooking sherry)

Melt the butter in a heavy 6-quart pot. Add the celery and onions and cook gently over low heat until translucent. Add the flour all at once and stir to blend well. Stir over low to medium heat to make a light, blond-colored roux. Slowly add the milk, then add the chicken base and white pepper. Cook for about 30 minutes, or until soup has thickened. Add shrimp and crawfish tails. Stir in sherry, blending well; cook for 5 minutes. Serve hot.

(Huisache Grill and Wine Bar)

GERMAN POTATO SOUP

This soup is a hands-down favorite at the New Braunfels Smokehouse. Through the years, it has warmed the souls of many weary travelers. The bits of dried beef add a rich, smoky taste.

Yield: 4 servings

3 tablespoons butter
½ cup chopped onion
4 large potatoes, peeled and cubed
2 teaspoons salt
1 teaspoon black pepper
7 cups water
3 ounces Smokehouse Dried Beef, shredded, or substitute 3 ounces cooked, chopped smoked bacon

Melt the butter in a heavy 6-quart soup pot. When butter has melted and the foam subsides, add the onions and sauté until wilted and transparent, about 5 minutes. Add the remaining ingredients except the dried beef. Bring to a boil, reduce heat, and simmer until potatoes are very soft, about 1 hour. Ladle the soup into bowls and top with the dried beef.

Variation: To make Vichyssoise, use 6 cups water and add 1 cup half-and-half and ¼ cup chopped chives to the recipe. Serve well chilled.

(New Braunfels Smokehouse)

New Braunfels Smokehouse

New Braunfels

The legendary New Braunfels Smokehouse started out as an ice plant where local farmers and ranchers brought their meats for cold storage. In 1943 San Antonio businessman and cattle rancher Russell Kemble "Kim" Dunbar bought the plant. Benno Schuenemann, the manager, had a talent for smoking meats and would use old German recipes to cure and smoke the customers' meats. His hickory-smoked hams, turkeys, and sausages were the talk of the entire area. ❧ As word spread, Kim Dunbar and his wife, Arabel, set up a mail-order business and then in 1952 opened a Tasting Room, inspired by old Smoky Mountain smokehouses with their bent stovepipes, leaning walls, and rustic furnishings. Thus the New Braunfels Smokehouse restaurant was born. ❧ In 1967, with the business growing, the original building was moved across the highway to its present location just off I-35 and expanded. The Yard, a shaded outdoor area that was opened in 1986, is the latest addition. ❧ The Dunbars' daughter and son-in-law, Susan and Dudley Snyder, now own and operate the New Braunfels Smokehouse. The restaurant serves breakfast, lunch, and dinner and is still a favorite stop with travelers. Its mail-order business, now expanded to include Internet sales, continues to grow.

140 State Hwy. 46 South (fronting I-35) • (830) 625-2416 • www.nbsmokehouse. com

Pipe Creek Junction Café

Pipe Creek

The Pipe Creek Junction Café is the antithesis of a tourist stop. Here, in the café's small dining room adorned with old advertising posters and furnished with vintage booths and tables and chairs, you'll find the locals gathered. The patrons stroll around from table to table, catching up on the latest gossip and news around town with their neighbors and friends. Owner Ginger Lee knows the customers by name.

❈ Ginger and her late husband, Don, former owners of the Helotes Café, purchased the building in 1987. It had served as the courtroom of the local justice of the peace as well as his general store and living quarters. The Lees refurbished the structure until the money ran out, and then they opened the café—ready or not! ❈ Over the years the café has earned a reputation for serving good, hearty, home-style foods and legendary pies and cobblers. *Texas Highways* listed the café first among its "Six Best Places in Texas for Fried Catfish and Shrimp." ❈ All-You-Can-Eat Shrimp Night brings customers from all over the region—and for good reason. Plates piled high with large, crispy fried shrimp and hand-cut fries emerge nonstop from the tiny kitchen with a side salad of crisp iceberg lettuce with tomato wedges and a package

of saltine crackers. Every meal comes with a bowl of fresh, seasonal cobbler with ice cream piled on top. ❈ Since Don's death in 2008, Ginger has continued to run this haven of comfort food as a legacy to his years of hard work and dedication. The café is open for lunch and dinner.

**9878 State Hwy. 16 South
(at FM 1283) •
(830) 535-6767**

BLACKBERRY COBBLER

Yield: 6 to 8 servings

2 cups flour
2 cups sugar
1 tablespoon baking powder
½ teaspoon salt
¾ cup whole milk
4 tablespoons butter
4 cups fresh blackberries, tossed with ¾ cup sugar and slightly mashed
Vanilla ice cream, if desired

Preheat oven to 350 degrees. Combine the flour, sugar, baking powder, and salt in a large bowl, whisking to blend well. Add the milk and stir until the mixture is well blended and has the consistency of pancake batter.

Melt the butter in a 13-by-9-inch baking pan. Pour the batter over the butter in the baking pan. Pour the blackberries over the batter, spreading evenly.

Bake in preheated oven for 1 hour, or until bubbly and batter is set. Serve warm, topped with vanilla ice cream, if desired.

(Pipe Creek Junction Café)

Cool Mint Café

San Marcos

Cool Mint Café opened its doors in 2006 in a 1920s Arts and Crafts bungalow adjacent to the Texas State University campus and the San Marcos Historic District. ❀ The goal of chef and owner Suzanne Perkins is to serve fresh, high-quality ingredients in interesting combinations and with skillful preparation and presentation. "We want our guests to leave the table satisfied and feeling better after they eat rather than feeling stuffed and dull by getting little or no nutritional value from the meal," she says. ❀ Cool Mint Café uses fresh, local organic vegetables and poultry and only fresh herbs, which come from Suzanne's extensive herb gardens. ❀ Armed with a background of managing resorts on California's Monterey Peninsula, Suzanne serves the community with a full café menu, an in-house bakery, and a take-out market offering fast service. Cool Mint also does catering and special-events planning. ❀ Diners are greeted with a tray of hot Texas yeast rolls made from scratch in the café's bakery, as are all of the desserts. The menu is simple, yet fun, and loaded with nutritious and delicious goodies. Appetizers range from a savory Mexican Cheesecake to an excellent Greek Platter—hummus served with olives, roasted peppers, and cucumber slices with toasted pita wedges. The Tomato-Basil Bisque is served with a mini grilled-cheese sandwich; heartier dishes include grilled entrées served with fresh sauces or salsas. Desserts from the bakery offer a selection of unique flavor pairings. ❀ The café composts all vegetable trimmings, and Suzanne continually strives to use ecologically beneficial practices while producing the finest-tasting, most nutritional foods possible. She's accomplished that goal handily. You'll never suspect that those fabulous tastes on the plate are actually good for you!

415 Burleson Street • (512) 396-2665 • www.coolmintcafe.com

SAN MARCOS

GARDEN GAZPACHO

Yield: 4 to 6 servings

6 medium tomatoes, seeded
2 medium cucumbers, peeled and seeded
Half of a large green bell pepper
Half of a large yellow bell pepper
1 medium sweet onion
¼ cup red wine vinegar
¼ cup extra-virgin olive oil
¼ cup Worcestershire sauce
2 tablespoons minced garlic chives
2 tablespoons minced fresh thyme
2 tablespoons minced fresh oregano
1 quart well-chilled tomato juice
2 cups ice cubes
Salt and pepper to taste
Additional chopped herbs for garnish
Lime slices for garnish

Chop the vegetables into medium dice. Combine in a large bowl and stir in the vinegar. Add olive oil, Worcestershire sauce, herbs, tomato juice, and ice cubes, blending well. Season with salt and pepper. Refrigerate until ready to serve. The ice will, of course, melt, so stir well just before serving. Garnish with additional herbs and a slice of lime per serving.

(Cool Mint Café)

G&R Grocery & Market

San Saba

San Saba is a charming small town at the northern boundary of the Hill Country. It's located adjacent to the San Saba River with its clear water and famous pecan bottoms. It's the county seat of San Saba County, so it's got a courthouse square with an array of shops and businesses. One of those shops in particular, right smack across the street from the courthouse, stands out. The G&R Grocery could have been lifted right out of Mayberry and plucked down there in San Saba. Except that the G&R Grocery has been around a lot longer than the TV Mayberry of my childhood. ❧ G&R Grocery was established over sixty-two years ago, in 1946. And Edward Ragsdale, who opened the market, is still its sole proprietor. It's the kind of store where I remember shopping as a kid with my grandparents on visits to their rural home. The well-worn wooden floors have a polished patina, and the original wooden bins and shelves hold canned goods and modern cleaning products. There are all the staples you need and a side room with charming old produce cases, perhaps the original ones, and, yep, still filled with fresh produce. ❧ The meats are fresh, arranged in an old meat case in the back of the store. I couldn't resist the temptation to purchase some of the best-looking pork chops I've seen in years. Thick-cut and rimmed with a half-inch ring of pork fat, just like the ones my mother used to fry up when we were kids, back in the good old days before we knew the evils that lurked in that delicious fat. I'd save the fat for the last and eat every bite of it! Where on earth did Mr. Ragsdale find those fat pigs? There were two-inch-thick rib eyes with incredible marbling and the biggest briskets I've ever seen. I tried to picture the truly awesome steers from which they must've come. Sausages and thick-sliced bacon were arranged in neat rows, next to fat hens,

rump roasts, and spare ribs. I was very much like a kid in a candy store standing there in front of all that beautiful meat! ❧ G&R Grocery was a divine discovery, a small-town treasure. If I lived in San Saba, I'd shop there. As it is, I have to make a long drive when I go back for more of that meat! ❧ Long live Edward Ragsdale and the G&R Grocery in San Saba, Texas!

502 E. Commerce Street • (325) 372-3364

The Great San Saba River Pecan Company

San Saba

San Saba is known as the Pecan Capital of Texas, and local folks take that designation seriously. So Larry and Martha Newkirk were elated when they had the opportunity to purchase an existing orchard of 10,000 pecan trees in 1988. Originally planted in 1968, the orchard had suffered from a few years of neglect, but the Newkirks slowly nourished the majestic trees back into shape for their commercial operation. ❦ The Great San Saba River Pecan Company is located in a beautiful parklike setting just west of town and bounded by four and a half miles of San Saba River frontage. It's a lovely place to visit, and the Newkirks have provided plenty of picnic tables where you can have a lovely picnic under the pecan trees. Martha's wall exhibit inside the charming retail shop tells the story of the area's pecan industry. ❦ The orchard is planted with the Western Schley variety, a very tasty paper-shell pecan and the most popular with commercial growers in the area. You can pick your own pecans during the season (November–January); old-fashioned cane poles are available for thrashing the pecans from the trees! Or you can purchase unshelled and shelled pecans by the pound at the retail shop or via the Web site. ❦ The shop also offers myriad pecan products made by the company, many packaged in attractive gift baskets. Your mouth will water at the free samples set up for you to taste around the store: fresh-brewed Pecan Coffee, various fruit and berry preserves combined with pecans, chocolate-covered pecans, and sweet and spicy seasoned pecans, or just plain, pristine pecan halves. Pecan Popcorn is a unique treat, and the Peach, Pecan & Amaretto Preserves won a coveted first prize in the Most Outstanding Preserve category at the International Fancy Food Show in New York. ❦ The Great San Saba River Pecan Company also produces baked goods: fruitcake, cinnamon-pecan streusel, pecan nut breads, and the award-winning Traditional Pecan Pie. One of the Newkirks' most unique products is Pecan Pie in a Jar—just pour the contents of the jar into a nine-inch pie shell, bake, and you've got yourself one fine pecan pie! ❦ All of the goodies produced by the company are made and shipped from facilities on the premises.

234 U.S. 190 West • (800) 621-9121 • www.greatpecans.com

TRADITIONAL PECAN PIE

Yield: One 9-inch pie

3 tablespoons butter
2 teaspoons vanilla extract
¾ cup sugar
3 eggs, well beaten
1 cup medium or large pecan pieces
1 cup light corn syrup
⅛ teaspoon salt
1 prepared unbaked 9-inch pie shell
½ cup pecan halves

Preheat oven to 450 degrees. Cream the butter and vanilla together, gradually adding the sugar. Add the eggs, beating well. Thoroughly blend in the pecan pieces, corn syrup, and salt. Turn the mixture out into the prepared pie shell. Bake in preheated oven for 10 minutes; reduce heat to 350 degrees, arrange the pecan halves over the top and bake an additional 30 to 35 minutes, or until the pie is set. Cool on wire rack before serving.

Note: Martha says you can make another excellent dessert—one with virtually no preparation—by using the Great San Saba River Pecan Company's Peach, Pecan & Amaretto Preserves: simply stir one jar of the preserves into a quart of softened vanilla ice cream, pour into a prepared graham cracker pie crust, and refreeze. Before serving, top with slices of fresh fruit and whipped cream.

(The Great San Saba River Pecan Company)

Oliver Pecan Company

San Saba

It was a late August afternoon when I first visited the 350-acre San Saba River "bottom" pecan orchard of the Oliver Pecan Company, known as its "home orchard." There was a certain earthy, slightly fecund smell to the land, rimmed by the meandering San Saba River, with the dense growth of towering native pecans overhead. On the ground there were patches of sunlight streaking through the trees. Wind rustled through the old branches, waving the heavy pecan clusters against a cerulean sky. ❧ Native Texas pecan trees were mentioned in Spanish explorer Cabeza de Vaca's report to the king of Spain in 1542. The tree has proven itself to be a survivor and is very resistant to drought, sending its taproot to astonishing depths to find water. The native pecan is tough nut to crack, however, with a very thick shell, making it a less attractive variety for commercial growers. But the intense flavor of the nut is preferred by fine bakeries. Native Texas pecans are the only pecans used by the Blue Bell Creamery in Brenham, makers of Blue Bell ice cream. ❧ Oliver Pecan Company began in 1970 when Gordon Lee and Clydene Oliver bought a piece of property on the San Saba River that had about 250 native pecan trees. To harvest the crop that fall, they used cane threshing poles and a lot of buckets. Eventually they established orchards with other varieties of pecan trees that are more commercially attractive than the native pecan. A few years later Gordon Lee was one of the first pecan growers in the Hill Country to use mechanical harvesting equipment. In 1980 they established a warehouse in Belmont to purchase pecans from other growers and broker them to major pecan-shelling companies. ❧ In 1984 the Olivers opened a warehouse and retail store in San Saba. Soon a shelling plant to process their own pecans followed, then a candy kitchen. Oliver Pecan makes over seventy-five different pecan-laden goodies in the candy kitchen. A warehouse was opened in San Angelo in 1992 to serve the West Texas growers, along with a custom shelling plant. ❧ Little did Gordon Lee and Clydene know that their 250 trees would be the beginning of a family pecan dynasty! Every child, spouse, and grandchild in the family is involved in the operation of the company in one capacity or another. The Olivers now have pecan orchards in San Saba, Gonzales, Luling, Belmont, Goldthwaite, and DeLeon, with a total of approximately 40,000 trees. Their operation also includes pecan retail stores in San Saba and San Angelo, a mail-order catalog, Web site, wholesale, and fundraising, in addition to a gourmet kitchen store in San Angelo (SugarBaker's), a gift shop in San Saba, a grain elevator, a fertilizer company, farming, and cattle. Gordon Lee passed away in 2000, but the company marches on under the direction of Clydene, sons Shawn and Mark, daughter Marcie Maxcey, and their spouses and children, all operating the family business.

1402 W. Wallace Street • (325) 372-5771 • www.oliverpecan.com

Sister Creek Vineyards
Sisterdale

Sister Creek Vineyards is located in the heart of the tiny Sisterdale community between the east and west forks of cypress-lined Sister Creek. Recently discovered historical documents indicate that vineyards were planted in this area in the 1860s. ❉ The winery, housed in a restored 1885 cotton gin, was established in 1988 by rancher Vernon Friesenhahn and his helper, Danny Hernandez, at a time when the rebirth of the Texas wine industry was in its infancy. Danny had gained an appreciation for wine in the 1970s while stationed in Germany, and the time-honored secrets that local winemakers shared with him proved to be beneficial when he became the winemaker at Sister Creek Vineyards. Twenty years later, he's still handcrafting each bottle with gusto, passion, and great joy. ❉ Danny uses classic French (Bordeaux and Burgundy) wine-making techniques to produce the Sister Creek wines. He ages the wines in sixty-gallon oak barrels for up to three years and uses minimal filtration to allow the wines to retain their fullest flavors. ❉ Sister Creek produces Chardonnay, Pinot Noir, Merlot, and superb Cabernet Sauvignon blends. One of its most noteworthy wines, and an excellent dessert wine, is a Muscat Canelli Reserve, a well-balanced wine that has intense fruit and a rich, long finish. It's a fun wine to which Danny adds just a hint of carbonation, making it spritz around the palate.

1142 Sisterdale Road • (830) 324-6704 • www.sistercreekvineyards.com

SISTERDALE

THE POODIE BURGER

Poodie Locke was the master of Hill Country bar food—and what better bar food is there than a really bodacious burger? The Poodie Burger ranks at the top of the list of best burgers around. Hatch chilies are essential to the fabulous taste of the Poodie Burger. This variety of Anaheim chili is grown only around Hatch, New Mexico, and is available only from late August to early September. When you find them in the supermarket, buy up tons of them, blister and peel them, then freeze them in small quantities so you can make Poodie Burgers 'til next year's crop.

Yield: Six 6-ounce patties

2¼ pounds ground Black Angus chuck, 80/20 meat-to-fat ratio
Salt and freshly ground black pepper
6 slices Monterey Jack cheese
Buttered and toasted burger buns
Hatch Chili Blend (see recipe below)
Grilled Onions (see recipe below)
Mustard and mayonnaise
Sliced hamburger pickles
Chopped raw onions
Iceberg lettuce leaves

Hatch Chili Blend

3 tablespoons butter
6 Hatch chilies, blistered, peeled, seeded, and cut into ½-inch dice
1 small onion, chopped
3 large garlic cloves, minced

Grilled Onions

1 large onion, sliced about ⅜-inch thick
Canola oil

Poodie's Hilltop Bar and Grill

Spicewood

To the fans of great Texas music, fabulous bar food, camaraderie, boot scootin', and, yes, beer and other libations of a stronger nature, Poodie's Hilltop Bar and Grill is a Hill Country legend. ❉ Now this ain't one of those places that puts an *e* on the end of "grill." It's an unassuming-looking place, one that you'll likely miss in the hustle and bustle of the present-day Highway 71 area between Oak Hill and Spicewood if you're not really looking for it. ❉ But don't be misled by the rather unremarkable exterior. Opened in 1998 by Randall "Poodie" Locke, Poodie's has long been a haven for bikers, musicians and wannabes, rockers young and old, aging hippies, schoolteachers, and button-down types. Inside you'll find the usual bar paraphernalia, but take a closer look and you'll find personalized photos of country music giants covering every inch of space not covered by a neon beer sign. Music memorabilia clutters every available nook and cranny. ❉ Before his death in 2009, Poodie Locke was Willie Nelson's road manager, a "day job" that he held down for over thirty years. Oh, the stories Poodie could tell! Every now and again, Willie, who lives right down the road, might just show up at Poodie's to throw back a brew or two and say howdy all around. ❉

The food at Poodie's is legendary in its own right. The Poodie Burger, with a side of hand-cut fries, is a masterpiece of decadently delicious, dripdown-the-front-of-your-shirt bar food. In addition to the comfort food, there's live music, dancing, pool, and shuffleboard—and a huge, two-level deck outside under the oak trees where you can watch a pretty special sunset. ❉ Imbued with Poodie's huge *joie de vivre*, the roadhouse continues to operate. If you squint, you just might catch his image at the end of the shuffleboard table.

22308 State Hwy. 71 West • (512) 264-0318 • www.poodies.info

Divide the ground chuck into six equal portions and pat into firm patties with no cracks or gaps. Season both sides with salt and a liberal amount of black pepper. Refrigerate until ready to grill.

Make the Hatch Chili Blend. Melt the butter in a heavy 10-inch skillet over medium-high heat. Add the chopped chilies, onion, and garlic. Cook, stirring, until onions are wilted, transparent, and lightly browned, about 7 minutes. Remove from heat and set aside; keep warm.

Grill the patties to the desired degree of doneness. Set patties to one side of the grill and top with a slice of cheese. Keep the patties on the grill just until the cheese begins to melt. While patties are grilling, liberally brush the sliced onions with canola oil and grill, turning only once, until the onions are slightly charred and limp. Remove from grill and set aside; keep warm.

To assemble the burgers, spread mustard on one side of each toasted bun and mayonnaise on the other side. Place a burger on the bottom half of each bun and top with a portion of Hatch Chili Blend. Add a slice or two of the Grilled Onions. Add pickle slices, chopped raw onions, and top with a couple of lettuce leaves. Add top bun and chow down!

(Poodie's Hilltop Bar and Grill)

Spicewood Vineyards

Spicewood

Spicewood Vineyards, one of the state's more established wineries, was founded by Edward and Madeleine Manigold in 1990 in southern Burnet County near Marble Falls. ❄ The winery was a retirement project for the Manigolds following long careers in public education. Passionate about making fine wines, they both studied enology and viticulture at the University of California at Davis and Grayson County College in Denison, Texas. So it was not by sheer luck that the winery is so successful, having received over fifty medals in every level of competition. ❄ The vineyard is located in an ancient bed of the Colorado River with a complex topsoil mixture, ranging from red clay loam to the Trinity Sands. The bedrock is limestone. The vines grow on a hillside with a peak elevation of 900 feet, utilizing a vertical-shoot-positioned trellis system to convert the maximum amount of Texas sun into fine wine. The first 1.5 acres of grapes were planted in 1992, with the acreage expanding gradually to a total of 17 acres in production by 1998. ❄ The original winery building resembles a traditional nineteenth-century Hill Country home with board-and-batten walls and a front porch. The new winery facility was completed in 1999, which provided the capacity to produce more than 5,000 cases per year and includes underground storage for more than 400 barrels. The tasting room, which overlooks the vineyards, has a covered pavilion shaded by oaks and has been the scene of numerous vintner's dinners, where area chefs prepare their finest offerings to be served with carefully paired varietals from Spicewood Vineyards. The pastoral setting has become a favorite with visitors. ❄ The wines from Spicewood Vineyards are handcrafted entirely from grapes grown on the estate. Major varietals include Cabernet Sauvignon, Merlot, Chardonnay, Zinfandel, Muscat Blanc, Semillon, Sauvignon Blanc, Cabernet Franc, and Johannisberg Riesling. Two of Spicewood's best varietals are an outstanding Sauvignon Blanc, the perfect Texas summer wine, and Semillon Reserve, a varietal that is often overlooked—but not this one: it's bold and flavor-packed. ❄ In the late summer of 2007, the Manigolds retired again, selling Spicewood Vineyards to Austin attorney and businessman Ron Yates. Now they can visit all those places they've heard about from travelers stopping by the tasting room. Ron plans to continue Ed and Madeleine's legacy of producing fine Texas wines from estate-grown grapes.

1419 Burnet County Road 409 • (830) 693-5328 • www.spicewoodvineyards.com

Becker Vineyards

Stonewall

Becker Vineyards, located three miles west of Stonewall, was established in 1992 by Richard and Bunny Becker. They planted thirty-six acres of French vinifera vines on a site where native mustang grapes had once grown. The original planting has been expanded to forty-two acres that produce eight varietals. ❋ A reproduction of a nineteenth-century German stone barn houses the winery and tasting room. Produced with

French methods, the wines are aged in oak in the largest underground wine cellar in Texas. With a fermenting capacity of 35,000 gallons, Becker Vineyards sells out of its wines every year. But the Beckers want to keep their operation small—they're interested in quality, not quantity. In the first half of 2007, the winery bottled 42,000 cases of wine representing twenty-two varietals! ❋ Becker wines have won innumerable medals and awards, and they've been served at the White House and at state functions on several occasions. New varietals continue to be produced. In 2007 Becker Vineyards released its first bottlings of Barbera and Malbec to rave reviews. ❋ The tasting room is an inviting space that has been furnished with an antique bar from San Antonio's Green Tree Saloon. The decor also features arts and crafts by local artisans displayed in antique pie safes and armoires from the surrounding area. ❋ Visitors touring the facility can stay in the rustic Homestead Bed and Breakfast, a renovated log cabin from 1890. Also on the grounds is the Lavender Haus Reception Hall, which is available for special events. ❋ With three acres behind the winery planted in lavender, Becker Vineyards hosts a very popular Lavender Festival each May. ❋ Becker wines and various lavender products are available through the Web site.

464 Becker Farms Road • (830) 644-2681 • www. beckervineyards.com

PORT WINE PECANS

These delectable little morsels are the creation of Paul Gingrich, who worked in the tasting room at Becker Vineyards for several years. Paul says they can be added to your favorite salad with a balsamic vinaigrette, sprinkled on ice cream, scattered over chocolate cake or cheesecake, crumbled over a warm wedge of Brie, or just munched like popcorn.

Yield: 2 cups

1 cup sugar
½ cup Becker Vineyards Vintage Port wine
2 cups pecan halves

Grease a heavy-duty baking sheet with solid shortening and set aside. Combine the sugar and wine in a heavy 14-inch skillet over medium-high heat. Stir the mixture constantly. When the sugar has dissolved and the syrupy mixture begins to boil, quickly stir in the pecans. Continue to stir, turning the pecan halves to prevent burning. Cook for about 5 to 6 minutes, or until the mixture has thickened and the syrup has turned a deep caramel color. Remove pan from heat and turn the pecans out onto the prepared baking sheet. Quickly separate the pecans into a single layer with a wooden spoon. (Take care not to touch the pecans; they are hot as blazes!) Allow pecans to cool for about 30 minutes, or until they are totally hardened. Store in a moisture-proof container.

(Paul Gingrich)

STONEWALL

137

ANCHO CHILI AND HONEY–BASTED QUAIL

Yield: 6 servings

12 semi-boneless quail
1 ancho chili, seeded and soaked in 1 cup of hot
  water for 30 minutes
1 garlic clove, peeled
2 tablespoons honey
1 teaspoon kosher salt

In a blender combine the ancho chili and soaking liquid, garlic, honey, and salt; puree until smooth.

Heat a gas grill to hot. Place the quail, breast side down, on the hot grill and cook for 3 to 4 minutes, or until skin is crisp. Turn over and baste with the ancho chili sauce. Cook for an additional 3 to 4 minutes. Move quail to cooler part of grill, turning them over so breast side is down and baste again. Keep turning and basting for 2 minutes more. Serve hot, basting again with the sauce.

(Mac & Ernie's Roadside Eatery)

Mac & Ernie's Roadside Eatery

Tarpley

Mac & Ernie's Roadside Eatery at the Williams Creek Depot in Tarpley is a rare find. Even rarer a find is the owner, Naylene Dillingham Stolzer. ❄ Naylene, raised in Boerne, has been cooking since she was seventeen, when she flipped burgers. She worked in various restaurants, learning the art of baking along the way. ❄ After Naylene married, she and her husband, Steve, started raising goats as a pastime, as did their neighbors, the McKinnerneys. To market the goat meat, the McKinnerneys suggested opening a small restaurant. It seemed like a good idea since Naylene had such an impressive restaurant background. ❄ They bought an 8-by-10-foot portable building, a five-burner gas grill, a roaster oven, and a used refrigerator, and they made a deal with the owners of Brown's Hilltop (now Williams Creek Depot), allowing them to set up shop on the edge of the store's parking lot. ❄ When Mac & Ernie's opened in 1999, the menu consisted of cabrito tacos, fajitas, sausage wraps, and beans. Business was good, so they bought a used fryer and started cooking catfish. The enterprise grew and the couples bought the present 10-by-16-foot building. The Stolzers bought out the McKinnerneys in 2000. ❄ Naylene began to expand the menu, adding items like lamb, salmon, and quail dishes. She added steaks as soon as she could afford steak knives. ❄ Today Naylene's upscale menu draws people from all over Texas to the tiny place. You buy your drinks (including beer and wine) at the Williams Creek Depot (they have a corkscrew at the counter), place your order at the window of Mac & Ernie's, and stake out a seat at one of the picnic tables. ❄ The dinner menu changes often and offers something for everyone—steak, quail, pork tenderloin, lamb chops, yellowfin tuna, fried catfish, fried shrimp. Everything comes with a huge, fresh salad with a tasty rosemary-balsamic dressing, and the sauces for the entrées are nothing shy of fabulous, all at reasonable prices. ❄ Mac & Ernie's is open for lunch on Wednesdays, Fridays, and Saturdays; for dinner on Fridays and Saturdays. Lunch is when you can get Naylene's legendary Cabrito Burger. ❄ Mac & Ernie's Roadside Eatery has been featured in *Texas Monthly*, *Southern Living*, and on the Food Network. Make the trip—it's a one-of-a-kind experience, and the food is positively to die for!

11999 FM 470 • (830) 562-3250 • www.macandernies.com

Fall Creek Vineyards

Tow

Located on the northwest shore of beautiful Lake Buchanan and a few miles north of the tiny hamlet of Tow, Fall Creek Vineyards is one of the most picturesque of the Hill Country wineries. It is situated on the Fall Creek Ranch, owned by Ed Auler's family for five generations. Fall Creek is one of the pioneer Texas wineries and the very first in the Hill Country. ❈ The winery began in 1975 as an ambitious project of Ed and his wife, Susan, to determine whether high-quality wine grapes could be grown on their ranch. The Aulers' curiosity for the project was sparked during a trip to the wine-growing regions of France, where they had intended to study cattle operations. They noticed that the soil, terrain, and microclimate of their ranch were remarkably similar to those in France. When they made the decision to give grape growing a try, they asked one of California's greatest winemakers, Andre Tchelistcheff, to come to Texas and have a look at their dirt. Andre determined that Fall Creek could grow notable red wine grape varietals from the Bordeaux region.

After tasting their first bottling from the Aulers' experimental vineyards, Tchelistcheff wrote to them with the admonition "Plant more!" ❈ The present winery facility was opened in the spring of 1983, surrounded by a sea of Texas bluebonnets. The ensuing years have seen the winery grow dramatically in size, quality, production, and recognition—Fall Creek has received too many accolades to list. ❈ Over the years the Aulers have maintained their cattle operations at the ranch while developing award-winning wines. They founded the

legendary Texas Hill Country Wine & Food Festival, hosting its first event in April of 1986. They have also been active in lobbying for state legislation instrumental to the success of the Texas wine industry. ❈ In 1990 the Aulers opened their present tasting room, an elegant facility with an inviting courtyard. It's a great place to sip some of the good wines available in the tasting room and enjoy the Hill Country. ❈ Fall Creek produces many varietals, including the super-premium wine Meritus, a blend of cabernet sauvignon, merlot, and malbec, first introduced in 1998. One of Fall Creek's best-selling wines is Caché, a blend of six white grapes that closely resembles the flavor of California's popular Caymus Conundrum. ❈ Chad Auler, Ed and Susan's son, has recently assumed the winery's sales and marketing operation, taking Fall Creek Vineyards into its second generation as a family business.

1820 CR 2241 • (325) 379-5361 • www.fcv.com

Certenberg Vineyard

Voca

As a young boy, Houston native Alphonse Dotson was fascinated by a massive grape arbor at his grandfather's house. He'd watch the grapes grow every year—big clusters of them hanging everywhere. For forty years Alphonse never forgot those grapes, even though he went on to become a star football player and eventually wound up as a defensive linebacker for the Oakland Raiders. ❄ He lived for some ten years in Acapulco, where he had met his wife, Martha, but in the early 1990s the pair decided to move to the United States. Then Alphonse remembered those grapes again. ❄ After talking to winemakers and viticulture experts in both California and Texas, Alphonse was hooked on the idea of growing grapes. He made dozens of charts, researched the subject to death, and finally went looking for land. He and Martha settled on a parcel in McCullough County with the red, sandy soil that he had been looking for. ❄ Alphonse named the vineyard after that grape-growing grandfather, Alphonse Certenberg. The Dotsons planted thirty acres in chardonnay, cabernet sauvignon, merlot, and muscat canelli grapes in 1997. The operation was totally hands-on at first because they simply weren't able to find help, but their hand-nurturing paid off in vigorous, healthy vines that produce high-quality wine grapes. ❄ Today Alphonse is affectionately referred to by fellow grape growers and winemakers as the state's "Celebrity Grape Grower." He was included on *Saveur* magazine's list of 100 Favorite Foods, Restaurants, Drinks, People, Places and Things. He served as the 2007–2008 president of the Texas Wine and Grape Growers Association. ❄ The Dotsons' dream is to open a small, very personal winery. ❄ The vineyard is not open to the public, but serious oenophiles may call to arrange a tour.

(325) 239-5500

Welfare Café & Biergarten
Welfare

The community of Welfare, situated between Waring and Boerne in west-central Kendall County and accessible by only a small county road, declined in population for most of the past century. The store and post office closed in 1978 and remained abandoned until 1998, when Gabriele Meissner McCormick and David Lawhorn opened the Welfare Café in the building. ❧ They've left the building much the way it was, and it's that rustic feel which adds to the ambiance of the place. There's outdoor dining in relaxing spaces in the back overlooking the scenic grounds. The Goat Barn, newly constructed by David from vintage materials, is a fabulous space for private parties. It feels like stepping back in time, perhaps to a dining and dance hall of the 1800s. ❧ But the draw that has made the Welfare Café a destination eatery for residents from all over the Hill Country and San Antonio is Gabrielle's food. The menu is a delightful, eclectic collection of very authentic German dishes, such as Rouladen (a stuffed filet of beef) and a selection of six different schnitzels, in addition to steaks and seafood. ❧ The wine list presents a nifty little collection of excellent labels and, of course, a lineup of great beers to enjoy in the Biergarten. Welfare Café is not a place where you stop for dinner on your way to the evening's entertainment—it is the evening's entertainment. The café is also open for lunch on Sundays.

233 Waring-Welfare Road • (830) 527-3700 • www.welfaretexas.com

CHICKEN FREDERICKSBURG

Chicken Fredericksburg is one of the most popular items on Gabrielle's menu, and she has graciously shared the recipe for the many fans of the dish.

Yield: 2 servings

2 tablespoons olive or canola oil

2 boneless, skinless chicken breasts, lightly floured

1 teaspoon minced garlic

1 teaspoon minced jalapeño, or to taste

¼ cup caramelized onions

1 ripe peach, peeled and sliced, with its juice

2 tablespoons dry white wine

½ cup whipping cream

1 tablespoon toasted pecan pieces

Spaetzle

2 eggs, beaten

½ cup milk

1½ cups all-purpose flour

¼ teaspoon salt

⅛ teaspoon freshly grated nutmeg

2 tablespoons butter

To make the spaetzle, combine the eggs and milk, then stir in flour, salt, and nutmeg, stirring until very smooth. Press the dough through a spaetzle maker or colander into boiling water. Boil for 2 to 3 minutes, then drain and set aside.

Prepare the chicken. Heat the oil in a heavy 10-inch skillet over medium heat. When oil is hot, sauté the chicken breasts on both sides until golden brown. Remove from pan, set aside, and keep warm. Add the garlic and jalapeño to the pan and cook until garlic is light tan. (Don't allow it to brown.) Add the onions and peach with its juice. Add the wine and deglaze the pan, scraping up browned bits from the bottom. Add the whipping cream. Return the chicken breasts to the pan and simmer until the sauce thickens, about 4 to 6 minutes.

While the sauce is thickening, finish cooking the spaetzle. Melt the butter in a heavy 10-inch skillet and sauté the spaetzle until light golden brown on the edges, about 3 to 4 minutes.

To serve, divide the spaetzle between two serving plates. Place a chicken breast on the spaetzle, top with a portion of the sauce, and garnish with toasted pecan pieces.

(Welfare Café & Biergarten)

Bella Vista Ranch
Wimberley

Olives in the Texas Hill Country may sound like a pipe dream, but the concept was actually the real dream of Jack Dougherty and his wife, Patricia. Modeled on the traditional Italian family farm, the property, a former goat farm, has over 1,000 producing olive trees. The main varieties are Mission and Coratina, with the Barouni and Pendolino varieties planted for pollination. The Doughertys have recently planted Arbequina, a popular Spanish variety. They also grow blackberries, raspberries, figs, and tomatoes. ✽ Marketing their olive oil under the Alfresco brand and their operation as the First Texas Olive Oil Company, the Doughertys use an authentic, Italian-made *frantoio* (olive press) designed to extract a high volume of oil by the cold-press method. The fall of 2001 saw the first crop, and production has increased each year. ✽ The ranch's tasting room offers tastings of Alfresco Extra Virgin Olive Oil along with wines produced from the grapes and berries grown in the vineyards. The estate-bottled olive oil, wines, vinegars, olive-oil products, jams and jellies, books, soaps, and olive trees are available for purchase at the tasting room and through the online gift shop. ✽ The Doughertys conduct tours of the orchard, and during the harvest season (usually September and October) you can even participate in the picking and pressing process that goes into the making of fine olive oil.

3101 Mount Sharp Road • (512) 847-6514 • www.bvranch.com

Kiss the Cook Kitchen Essentials

Wimberley and Boerne

After owning a cookware store in the West Texas town of Abilene, partners and best friends Janet Galloway and Bren Kirschbaum relocated to Wimberley and opened Kiss the Cook in a little building on the downtown square in 2001. The pair opened a second store in Boerne in 2006 in a newly renovated old house right off Main Street in the quaint downtown shopping district.

❋ Both stores are stocked to the ceiling with a vast selection of kitchen gear that makes a serious cook's heart race. Customers from far and wide love shopping for the very finest cookware, kitchen gadgets, bakeware, small appliances, cutlery, linens, serving pieces—you name it. If it belongs in a kitchen, Bren and Janet have it! Furthermore, they can tell you how to use it. Hospitality and customer service are their top priorities. This philosophy has paid off, garnering Kiss the Cook recognition from *Southern Living* as a Southern Living Favorite in 2003 and one of the magazine's Fifty Top Shops in 2006. ❋ In the fall of 2007 the ambitious pair realized their ultimate dream of opening a cooking school in a lovely new facility adjacent to the Boerne store. The classroom is equipped with the most up-to-date appliances and stocked with the fine cookware and gadgets available in the stores. Cookbook authors and chefs taught the first series of classes, which were eagerly attended, often with waiting lists. ❋ The shop's Web site is a great resource for cooks, too, with all sorts of cooking tips and information, tons of recipes, and the opportunity to sign up for the newsletter. If you're a serious cook, then you'll want to make Kiss the Cook a destination stop.

201 Wimberley Square (Wimberley) •
(512) 847-1553

113 E. Theissen Street (Boerne) •
(830) 249-3637 • www.kissthecooktx.com

Texas Specialty Cut Flowers
Wimberley

The bounty of gorgeous cut-flower arrangements from the Arnosky Family Farm has been providing bursts of welcome color and greenery to Hill Country homes since 1990. It's been a hard labor of love. ✤ On the farm's original twelve acres Frank and Pamela Arnosky built the first greenhouse and their bright blue home. When this acreage was in operation, they bought the adjoining eighteen acres. In 2003 they bought another one hundred acres on the site of the old Payton Colony, established in the 1870s as a free slave community. The former residents were tomato growers who took their tomatoes by mule cart to the tomato packing sheds at nearby Henly. The Arnoskys were able to buy the land only on the condition that they would continue to farm it. ✤ Today the Arnosky Family Farm has over forty acres in cultivation and fourteen greenhouses. In 2006, with the help of over 200 volunteers, the Arnoskys held an old-fashioned barn-raising at the Payton Colony site. Dubbed the Big Blue Barn, the structure houses the Farm Market, where you can purchase fresh-cut flowers, Pamela's gorgeous handmade bouquets, fresh-picked vegetables, locally made cheeses, fresh herbs, and plants. More than sixty kinds of flowers grown by Texas Specialty Cut Flowers are also sold at Central Market, Whole Foods Market, and select H-E-B stores in Texas as well as Farm to Market Grocery in Austin, King Feed in Wimberley, and McCall Creek Farms Market in Blanco. ✤

The farm is truly a family operation, with each of the four Arnosky children having specific chores along with raising chickens, milk goats, rabbits, and horses. The Arnoskys believe in environmentally sound farming practices, using only organic compost as fertilizer, and are dedicated to supporting the return of the small family farm.

**12550 FM 2325 (at FM 165) • (830) 833-5428 •
www.texascolor.com**

LAVENDER LEMONADE

Yield: 1½ quarts

4 cups plus 1 cup water
¼ cup chopped fresh lavender leaves
1 cup sugar, or more to taste
1 cup lemon juice (about 6 large lemons), or more
  to taste
Lavender bloom stems for garnish

Bring 1 cup of the water to a boil in a medium saucepan.
Place the lavender leaves in a medium bowl and pour the
boiling water over the leaves; cover tightly with plastic wrap
and allow to steep for 10 minutes. Strain through a fine
strainer into a clean bowl; set aside.

Combine the remaining 4 cups of water and sugar in
saucepan. Bring the mixture to a boil and boil for 3 minutes,
or until the sugar is completely dissolved. Do not stir.
Remove from heat and pour the sugar syrup into a large 2-
quart pitcher. Add the lavender water and lemon juice, stirring to blend well. Cover and refrigerate until well chilled.
Serve over ice, garnished with a blooming lavender stem.

(Wimberley Lavender Farm)

Wimberley Lavender Farm
Wimberley

Wimberley Lavender
Farm is owned by Karen
and O'Neil Provost, who
planted 3,000 lavender
plants of the Provence
variety in the fall of
2004. This variety has
a very high oil content,
a sweet scent, and low
levels of camphor, mak-
ing it ideal for culinary
use and bouquets. They
have gradually added
other varieties: Grosso,
Buena Vista, Munstead,
Twickle Purple, Alba (a white lavender), and Spanish lavender. ❀ One the most unique features at
Wimberley Lavender Farm is the Labyrinth, which consists of more than 600 lavender plants that
form concentric half-circles based on an ancient pattern inlaid in the floor of Chartres Cathedral
in France. Visitors can follow the twists and turns in and around the rings into the center in what
the Provosts hope will be an interesting, meditative journey, which takes about fifteen minutes
and ends with an expansive Hill Country view. ❀ During the lavender season a cozy shop on the
farm is stocked with lavender luxuries, such as handmade soaps, oils, candles, and personal care

products. Visitors
can also cut their
own lavender dur-
ing the season and
taste Karen's deli-
cious, very refresh-
ing Lavender
Lemonade.

**11300 FM 2325 •
(830) 833-1595 •
www.wimberley-
lavender.com**

Wimberley Pie Company

Wimberley

Neal Mallard, the owner of Wimberley's beloved Wimberley Pie Company, is a very serious guy. That's probably why he's such a successful baker, even though it was a craft he learned somewhat late in life. Neal, a music major in college and still a passionate music lover, plays the trombone in several local groups and is also a ringer in the handbell choir at his church. Even his hobbies sound like a full-time job, but there's more . . . . ❄ Neal knew nothing about baking in 1989 when he bought the business, a wholesale operation that baked pies and cheesecakes for restaurants. But he jumped right in and learned the ropes—from Hobart mixers to Blodgett pizza ovens, where he bakes seventy-eight pies at a time! Neal does all the baking himself and has only one full-time employee. ❄ Wimberley Pie Company is now strictly retail and ships pies all over the country through phone orders in addition to operating a storefront where pies can be bought whole or enjoyed in the small dining room by the slice. The company now does as much business in November and December as it did in an entire year when Neal bought it. ❄ You won't believe the amazing array of pies available— from the ever-popular Apple and Cherry to seasonal favorites like Pumpkin to more unusual creations like Fudge Pecan and Peanut Butter. He even has pies we don't often see in the Hill Country—like Rhubarb-Strawberry. Cheesecakes, quiches, and cookies are also available ❄ Serious pies from a serious guy!

13619 RR 12 • (512) 847-9462 • www.wimberleypie.com

KEY LIME PIE

Yield: One 10-inch pie

Crust

- 1¼ cups finely ground graham cracker crumbs
- 2 tablespoons plus 1½ teaspoons granulated sugar
- ¼ cup melted butter

Filling

- 3 egg yolks
- 8 ounces cream cheese at room temperature
- 1 (14-ounce) can sweetened condensed milk
- ½ cup fresh lime juice or key lime juice
- Sweetened whipped cream for garnish, if desired

Preheat oven to 350 degrees. Mix the graham cracker crumbs and sugar together, tossing to blend. Stir in the melted butter, coating well. Turn the mixture out into a 10-inch pie pan. Using the back of a spoon, press mixture against the sides of the pan to a thickness of ¼ inch, taking care to follow the slope of the pan. Spread the remaining mixture in the bottom of the pan and press down with a flat surface like the bottom of a measuring cup. Bake in preheated oven for 10 minutes. Remove from oven and allow to cool. (The crust can also be frozen for later use.)

While the crust is cooling, make the filling. Place the egg yolks in bowl of electric stand mixer and beat until light lemon yellow in color and very thick and "ribbonlike" in texture, about 5 minutes. Using a rubber spatula, scrape the thickened yolks into a separate bowl and set aside. Put the cream cheese in the mixer bowl and beat at medium speed until light and fluffy, scraping down the sides of the bowl often.

Return the yolks to the mixer bowl and beat at high speed for 3 minutes, scraping the sides of the bowl as needed. Add the sweetened condensed milk and mix at high speed for 3 minutes, scraping the bowl as needed. Add the lime juice and beat at medium speed for 3 minutes, scraping the bowl as needed. The mixture should begin to thicken slightly as you beat in the lime juice.

Pour the filling into the cooled pie crust and bake in 350 degree oven for 7 minutes. Allow the pie to rest for 30 minutes at room temperature, then place in an airtight container or carefully wrap in plastic wrap and refrigerate overnight before slicing. (This pie freezes quite well.)

Before serving, top each slice with a generous dollop of sweetened whipped cream, if desired.

(Wimberley Pie Company)

Acknowledgments

With each book project it always seems so difficult to know where to start in thanking all of the many people who helped to make it happen.

First, of course, there are the folks who are featured in this book. They generously arranged, and often rearranged, their schedules to accommodate our long itinerary, which was so often interrupted by the record rains of the spring and summer of 2007. Then there were my many e-mails and phone calls requesting clarification or just one more quote. I hope we've done you all the justice you so richly deserve for your contributions toward making the Texas Hill Country such a special place.

It was such an immense joy to work with my sister Sandy Wilson, photographer extraordinaire, for the first time on a book. Her stunning photos bring my words to life, allowing the reader to identify with the people and really see and feel all of the bounty of the Texas Hill Country featured in this book.

Many thanks are in order to my publisher, Kathy Shearer, for giving me the opportunity to do this book, especially knowing it's one she's wanted to publish for a long time. Kudos to Alison Tartt for her unparalleled editing skills, to Barbara Jezek for yet another superb job of book design, and to the entire team at Shearer Publishing for their hard work behind the scenes.

My dear bear of a husband, Roger, survived yet another book, lending his unfailing support as dinner cook, foot massager, and bedtime collaborator as I would read segments of the manuscript aloud in bed as I edited. His critiques and comments would often provide just the perfect phrase for which I was racking my brain.

And, as always, a profound thank you to those professional friends and mentors who helped me become who and what I am professionally—Nathalie Dupree, Shirley Corriher, Paul Prudhomme, Francois Dionot, Blake Swihart, Rose Beranbaum, and three who have left us: Lee Barnes, Bert Greene, and Sharon Tyler-Herbst.

Terry Thompson-Anderson
Fredericksburg, February 2008

A book like this cannot be undertaken without the support and interest of more people than I can imagine, and I am extremely grateful to each of them. Of course, there is my sister and author Terry Thompson-Anderson and publisher Kathy Shearer, without whom I would never have had the opportunity to share in all these experiences. There is Barbara Jezek as well, the designer with whom I have worked over many years and whose book design can enhance any photographs.

My sister was unrelenting in her research of a complex assortment of people and places. She managed to schedule access to restaurants, kitchens, farms, retail shops, breweries, wineries, and people, all of whom were very generous with their time, throughout a large geographical area with the least amount of driving possible. And I cannot forget Roger Anderson, Terry's husband, who often had dinner on the table when we arrived, exhausted after a long day of driving, photographing, and interviewing.

I am very grateful to my friend Rosemary Schouten for educating me about both food and wine and their connection with the soil and people, without which I would have been unable to appreciate all the tasty things, beautiful places, and fascinating people I was photographing.

A very special thank you goes to my husband, Steve Freeman, who not only endured my prolonged absences but flew me back and forth in his plane several times for "just one more photo" in order to save me long hours on the road.

Sandy Wilson
Bellaire, February 2008

Index